MICROBES MINDCROBES

Human Entanglement with Microbes
On a Physical, Mental, Emotional
And Quantum Level

Helga Zelinski PhD

Microbes Mindcrobes

ISBN 10 - 0991296419
ISBN 13 - 978-0-9912964-1-5

Published by: Expert Author Publishing
http://expertauthorpublishing.com

Canadian Address:
1265 Charter Hill Drive
Coquitlam, BC, V3E 1P1
Phone: (604) 941-3041
Fax: (604) 944-7993

US Address:
1300 Boblett Street
Unit A-218
Blaine, WA 98230
Phone: (866) 492-6623
Fax: (250) 493-6603

TABLE OF CONTENTS

INTRODUCTION

Microbes were the first forms of life on this planet and have survived by adapting to ever changing environments, from simple one-celled life forms to intelligent, decision making, life sustaining species in charge of many primary functions in Earths bio-chemical and biological balances.

The scientific community estimates that life began approximately 3.5 billion years ago as a result of a complex sequence of chemical reactions that took place in the Earth's atmosphere. There was virtually no oxygen and these first micro-organisms were surviving by eating naturally occurring foods.

Gradual changes to these earliest cells resulted in new life forms which were no longer dependent on the same food supply as their ancestors, they were able to feed themselves by using the energy of the sun.

Without the activity of these early organisms, the Earth's atmosphere would still be without oxygen and the evolution of oxygen-dependent animals, including humans would have never occurred.

Micro-organisms are found in every environment from the deepest sea to the highest mountains and from the deserts to the poles. Microbes are in the air we breathe, the water we drink and the food we eat. They are also found in the soil, in plants, animals . . . and the human body.

The number of bacteria living within the human body of the average human adult are estimated to outnumber human cells TEN to ONE and are found mostly on our skin, the respiratory tract, the digestive system and the oral cavity. Microbes control every aspect of our lives

Exposure to bacteria and/or viruses and our interaction with these invaders will largely depend on the health of our internal environment and our mental/emotional state.

In order to understand how changes in bacterial populations affect us we must consider lifestyle, nutrition, personal hygiene, exposure to stress, pollution and the environment.

Many single organisms exhibit intelligence of a kind not seen in other species of the animal or plant kingdom. They neither have nervous systems nor brains, but harbor an internal system that can be equated to a biological computer.

To solve newly encountered problems, they assess the situation, recall stored data of past experiences, and then execute information processing, transforming the colony into a super brain.

Bacteria do not just react to change in their surroundings, they anticipate and prepare for it. They are not simple solitary organisms, they are highly social and evolved creatures. They congregate to fend off enemies, meet challenges of nature to reproduce, obtain food and maintain their critical environment

Some bacterial intelligence if compared to human levels is 60 points higher than the human average of an IQ of 100.

Microbes can keep us healthy and fend off invaders or make us very ill and may kill us under the right circumstances. We must provide a healthy environment for our resident bacteria to flourish and to help us maintain physical, mental and emotional health.

Regenerating our individual bio-terrain means forming alliances, not antagonisms with the microbial community.

Chapter 1
IN THE BEGINNING

How did Life Begin on Earth?

There are at least three theories which attempt to explain the origin of life on Earth. The first and oldest of these suggests that life was created by a Supreme Being or Spiritual Force. Most cultures and religions have their own explanations of creation that are passed down from generation to generation. Because these ideas cannot be proved or disproved, we consider them outside the boundaries of science. For that reason, they will not be pursued here and are left to each individual to decide.

The second theory suggests that life began somewhere else, in another part of the universe and arrived on Earth by chance, such as with the crash of a comet or meteor.

The third, and most common theory in the scientific community, is that life began approximately 3.5 billion years ago as the result of a complex sequence of chemical reactions that took place spontaneously in Earth's atmosphere. Based on these assumptions we examine early life from the theory of evolution.

Earliest Forms of Life

No one knows exactly when or how life began and the final, most important events leading to the origin of life are perhaps the least understood. Early Earth was dominated by volcanoes, a gray lifeless ocean and a turbulent atmosphere. Vigorous chemical activity occurred in heavy clouds, which were fed by volcanoes and penetrated both by lightening discharges and solar radiation.

The ocean received organic matter from the land and the atmosphere, as well as from in-falling meteorites and comets. Here, substances such as water, carbon, dioxide, methane and hydrogen cyanide formed key molecules such as sugars, amino acids and nucleotides. Such molecules are the building blocks of protein and nucleic acids, compounds ubiquitous to all living organism.

The first beings were probably much like tiny spherical droplet (coacervates) of assorted organic molecules – specifically, lipid molecules, which are held together by hydrophobic forces from a surrounding liquid. As a group, these bacteria are known as heterotrophic anaerobes.

Because there was virtually no oxygen in the atmosphere at this time, these bacteria were necessary anaerobic, meaning they did not breathe oxygen. Heterotrophs, meaning other feeders, are simply organisms that cannot make their own food. So heterotrophic anaerobes means they were creatures which ate some naturally occurring food and did not breathe oxygen.

The fossils of some of these oldest known forms

of life have been found in Australian rocks dating back 3.5 billion years. Most bacteria, which is generally accepted to be the first viable life form, contains several thousand genes, and is made up of hundred thousand millions of atoms, all linked and functional.

To create energy, these early bacteria probably consumed naturally occurring amino acids. Amino acids, sugars, and other organic compounds formed spontaneously in the atmosphere then dissolved in water. Upon digesting these molecules, early bacteria produced methane and carbon dioxide as waste products. Fermenting bacteria would be an example of what these early creatures might have been like.

Although the beginnings of life were successfully reproducing, their food sources (the organic matters) would not be able to sustain life indefinitely. In light of this, the organisms on Earth at that time would have to diversify over the long term to survive.

Evolution and Diversity

Gradual changes in the earliest cells gave rise to new life forms. These new cells were very different from the earlier heterotrophs because they were able to get their energy from a new source – **the Sun**.

And because these bacteria were able to feed themselves by using the energy of the sun, they were no longer dependent on the same food supply as their ancestors and were able to flourish. Photosynthesis is the process by which autotrophs (self-feeders) convert water, carbon dioxide, and solar energy into sugars and oxygen.

Eventually, photosynthesis by the earliest forms

of plant life (a form of life capable of feeding itself instead of feeding off others) began to produce significant amounts of oxygen. This had given life on Earth a whole new energy resource, allowing evolution to take place.

Over millions of years of evolution, photo-synthetic bacteria eventually gave rise to modern day plants. The appearance of organisms capable of performing photosynthesis was very significant – without this activity of these early bacteria, Earth's atmosphere would still be **without oxygen and the appearance of oxygen-dependent animals, including humans, would never have occurred!**

A critical early triumph was the development of RNA and DNA molecules, which directed biological processes and preserved life's operating instructions for future generations. But the origin of life was triggered not only by special molecules such as RNA and DNA, but also by the chemical and physical properties of Earth's primitive environments.

Most of life's history involved the biochemical evolution of single-celled microorganisms. We find individual fossilized microbes in rocks 3.5 billion years old, yet we can only conclusively identify multi-celled fossils in rocks younger than 1 billion years.

The oldest microbial communities often constructed layered mound-shaped deposits called stromatolites, whose structures suggest that those organisms sought light and were therefore pho-to-synthetic. These early stromatolites grew along ancient seacoasts and endured harsh sunlight as well as episodic wetting and drying by the tides. Thus it appears that micro-organisms had become remarkably durable and sophisticated.

Many important events marked the interval between 1 billion and 3 billion years ago. Smaller strips of land dominated by volcanoes were joined by larger, more stable continents. Life learned how to extract oxygen from water, and living things populated the newly expanded continental shelf regions. Between 1 billion and 2 billion years ago, eukaryotic cells (those with a nucleus) developed, with complex systems of organelles and membranes. These organisms then began to experiment with multi-celled body structures.

The discovery of fossilized filaments from bacteria or blue-green algae indicates that land was continuously vegetated between 1,200 and 800 million years ago, much earlier than was previously thought. The evolution of the plants and animals most familiar to us occurred only in the last 550 million years.

Marine invertebrates (such as shell-making ammonites) appeared first, then fish, amphibians, reptiles, birds, mammals and humans. Land plant communities also evolved from relatively ancient club mosses, horsetails and ferns, to the more recent gymnosperms (for example, conifers) and angiosperms (flowering plants).

Human Ancestors

The closest living relatives to humans are baboons, chimpanzees and gorillas. With the sequencing of both the human and chimpanzee genome, current estimates of the similarity between their DNA sequences range between 95% and 99%. By using a technique called the molecular clock which estimates the time required for the number of mutations to accumulate between two lineages, the approximate date for the split can be calculated.

Some species of the early hominins adapted to the drier environments outside the equatorial belt, along with antelopes, hyenas, dogs, pigs, elephants, and horses. The equatorial belt contracted about 8 million years ago. There is very little fossil evidence of the separation of the hominine lineage from the lineages of gorillas and chimpanzees.

By comparing the parts of the genome that are not under natural selection and which therefore accumulate mutations at a fairly steady rate, it is possible to reconstruct a genetic tree incorporating the entire human species since the last shared ancestor.

Each time a certain mutation appears in an individual and is passed on to his or her descendants a haplogroup is formed including all of the descendants of the individual who will also carry that mutation.

Adam and Eve

By comparing mitochondrial DNA which is inherited only from the mother, geneticists have concluded that the last female common ancestor whose genetic marker is found in all modern humans, the so-called, mitochondrial Eve, must have lived around 200,000 years ago.

The out of Africa model proposed that modern

Homo sapiens originated in Africa and migrated to Eurasia resulting in complete replacement of all other Homo species. Analysis have shown a greater diversity of DNA patterns throughout Africa, consistent with the idea that Africa is the **ancestral home of "the mitochondrial Eve and the Y-chromosome Adam".**

This theory predicts that all mitochondrial genomes today should be traceable to a single woman, the "mitochondrial Eve". **Whereas the Y chromosome is passed from father to son, mitochondrial DNA (mtDNA) is passed from mother to daughter and son.**

Genetic data can provide important insight into human evolution and studies show how one human genome differs from the other, the evolutionary past and its current effects. Differences between genomes have anthropological, medical and forensic implications and applications.

The earliest members of the genus *Homo* are Homo hablis which evolved around 2.3 million years ago and are the first species for which we have positive evidence of the use of stone tools.

Some scientists consider a larger bodied group of fossils with similar morphology to the original fossils, to be a separate species while others consider them to simply representing species internal variation, or perhaps even sexual dimorphism, meaning the development of the male and female differences in shape and appearance.

Homo habilis were about the same size as that of a chimpanzee, but their main contribution was bipedalism, walking on two legs as an adaptation to terrestrial living.

Scientists revealed their recent discovery of a new species of a human ancestor, dubbed Australopithecus garhi. The cranial and tooth remains are estimated to be approximately 2.5 million years old. These hominids provide insight into a crucial period of human evolution some two to three million years ago.

The anatomical evidence from garhi presents a sharp differentiation from A.africanus and provides starling evidence that modern humans may have later branched from garhi. The researchers discovered other remains, which would indicate, that the hominids walked on legs similar to modern humans and used rudimentary tools to strip away animal flesh.

Interestingly, the famous Lucy had long arms in comparison to her legs, while H.erectus had the proportions of modern humans. The proportions of the unidentified species were between the two. This suggests that the femur (thigh bone) extended before the forearm condensed.

Early human ancestors seem to have taken different climates and vegetation types in stride as they evolved from primitive populations in Africa to a worldwide, highly diverse human species.

Adaptation to Survive

A component of all existing life is that it adapts to survive. First it has to be able to reproduce, than it has to present a variation, so that all of the new generation is not identical either to the previous generation or to all its related species.

Once that variation is established, natural selection can take place by either differential birth or death adaptation. If species did not have the instinctive nature to survive via their genetic information, there would be no desire to adapt and evolve.

Chapter 2
AGRICULTURE

Living with Nature

Ten thousand years ago we were all Hunter-Gatherers, living with Nature. Then came the Horticulturist, (the Shaman, medicine woman and herb gatherers), they knew the healing and destructive capabilities of plants. The Horticulturist realized certain plants could be domesticated such as wheat and began to appropriate land to grow these plants. Towns and villages began to be populated because food could be grown and water could be found easily.

Although Agriculture was developed at least 10.000 years ago, it has undergone significant developments since the time of the earliest cultivation. Evidence points to the Fertile Crescent of the Middle East as the site of the earliest planned sowing and harvesting of plants that had previously been gathered in the wild. Independent development of agriculture is also believed to have occurred in northern and southern China, Africa's Sahel, New Guinea and several regions of the Americas.

Practices such as irrigation, crop rotation, fertilizers, and pesticides were developed long ago but have made great strides in the past century. The Haber-Bosch process (a method for synthesizing ammonium nitrate) represented a major breakthrough and allowed crop yield to overcome previous constraints. An intimate understanding of the local ecology is necessary for successful agriculture, and it may be important to extend this knowledge to the smallest of life forms. It is important to recognize that soils can vary tremendously as to the type and numbers of micro-organisms.

These can be both beneficial and harmful to plants and often the predominance of either one depends on the cultural management practices that are applied. It should also be emphasized that most fertile and productive soils have a high content of organic matter and generally, have large populations of highly diverse micro-organisms. Probiotics provide sustainable options for improved agricultural and environmental performance.

Soil, Plants and Trees

All living systems – including soil, plants, and trees – have a microbial ecology that can be managed and improved by the constant delivery of probiotics. Regenerating good bacteria produces a microbial ecology where beneficial bacteria dominate harmful bacteria, creating a healthier, more vibrant environment.

A cup full of healthy soil contains more microbes than there are people in the world. When we feed the microbes in the soil, they in turn feed the plants,

they look for water, air and organic matter to generate food a plant needs to do well. All agriculture is based on the mighty microbe. Weeds, insects and disease are only indicators of malnutrition in the soil and the best use of our understanding of Nature can be in how we grow our food and plants in healthy soils and prepare the way for future generations of humans to enjoy our planet Earth.

Chapter 3
MICROBES

The Oldest Living Organism

A microbe or micro-organism is an organism that is microscopic (too small to be seen by the naked eye), mostly unicellular or lives in a colony of cellular organisms. They are very diverse and include bacteria, fungi, archaea and protists, microscopic plants (green algae), and animals such as plankton and the planarian. Microbiologists also include viruses, but they are considered as non-living.

Single-celled micro-organisms were the first forms of life to develop on Earth and can be found in almost every habitat present in nature. Even in hostile environments such as poles, deserts, rocks, acid lakes and the deep sea. Microorganisms, by their omnipresence, impact the entire biosphere.

Microbes, especially bacteria, often engage in intimate symbiotic relationships (either positive or negative) with other larger organisms; some of which are mutually beneficial (mutualism), while others can be damaging to the host (parasitism). If micro-organisms can cause disease in a host they are known as pathogens. Pathogenic microbes are harmful, since they invade and grow within other organisms, causing diseases that kill humans, animals and plants.

Micro-organisms are vital to humans and the environment, as they participate in the Earth's element cycles such as carbon and nitrogen cycles, as well fulfilling other vital roles in virtually all ecosystems, such as recycling other organisms' dead remains and waste products through decomposition back to its natural state.

Although each individual microbe is but an almost weightless, one-celled organism, **microbes account for most of the planet's biomass – the total weight of all living things. It is no understatement to say that the microbes run the world**.

With their mighty collective strength, microbes control every ecological process, from the decay of dead plants and animals to the production of oxygen. No part of Earth escapes the influence of microbes, the oldest living organism.

Since microbes first appeared 3.5 billion years ago, they have diversified enough to colonize every eco-system, from scalding vents at the bottom of the ocean to burning desert sands to polar ice.

Shapes and Sizes

Microbes come in many different shapes and sizes. Common shapes are rod-shaped (bacillus), sphere-shaped (coccus) and helix-shaped (spirilla). These shapes are caused by the growth of the cell wall of the bacterium and usually acts to protect the microbe against invasion by other organisms or by chemicals. However, sometimes, in particular environments, microbes can exist without cell walls.

Bacteria are single-celled microorganisms, they reproduce by what is known as vegetative/asexual reproduction. This means that each cell divides into new, genetically identical cells, with each new cell functioning as an independent unit.

This process is known as binary fission and each cell will continue to multiply until the growth condition diminishes. The time it takes to accumulate all necessary components to divide is known as the generation length, which varies greatly between different genus and species, from as short as twenty minutes for E.coli to as long as twenty-four hours for Mycobacterium tuberculosis.

The population growth curve for bacteria is an exponential curve, whereby with each generation the number of bacteria doubles and under ideal circumstances (i.e. a ready supply of nutrients and a benign environment), a single E.coli bacterium can grow to more than one million bacteria in as little as three and a half hours. For one Mycobacterium tuberculosis bacterium to generate the same number, again under ideal circumstances, may take as long as ten days.

Every bacterium has a set of genes that completely describes and dictates its physical, chemical and biological characteristics. These genes are made of DNA and RNA and are known as the genotype of the bacterium. Usually, when a parent bacterium splits into two, the two progeny bacteria are genetically identical, i.e. they have the same genotype.

However, this is not always the case. There are several situations in which the genes (genotype) of a bacteria can change.

Mutation – Mutation happens when there is a genetic error in the copying of the genes from the parent to progeny bacterium. This results in a progeny bacterium that has a different genotype to that of its parent. Mutation rates vary between different genus and species of bacteria. Statistically, random mutations may occur as often as one in every million multiplications, or as seldom as one in every billion multiplications. However, since most bacterial population in the human body number well into the millions, if not billions the chances are that there will be many mutations with each generation.

Transduction – Bacteria, like humans, can be attacked by viruses. These bacterial viruses are known as bacteriophages, they invade bacteria and can change their DNA or may carry DNA from one bacterium to another and thereby alter the genotype of the bacterium – this process is known as transduction.

Conjugation – Sometimes bacteria may join together and exchange DNA. This process is known as conjugation and changes the genotype of the bacteria. Why are the above important? Because they allow the bacteria to adapt to their environment. Changes in the genotype may allow the bacteria to obtain nutrition from sources they were unable to feed from before, they may allow bacteria to survive in a more hostile environment, and they may allow the bacteria to avoid the action of destructive chemicals (e.g. antibiotics) or allow them to produce chemicals that protect from attack by organisms that are capable of destroying them.

Microbes can form an endo-symbiotic relationship with other, larger organisms, for example, **the bacteria that live within the human digestive system contribute to gut immunity, synthesize vitamins such as folic acid and biotin, and ferment complex indigestible carbohydrates.**

We may not realize it, but each one of us is a walking ecosystem. Most of the time we share our bodies harmoniously with about 90 trillion or so microbes, which in a healthy adult are estimated to **outnumber human cells by a factor of ten to one**. Living with microbes demands a biological balancing act and for the most part, though, we are blissfully oblivious to the microscopic life we carry around with us.

Microbes are in the food we eat, the air we breathe and the water we drink. **If all of the Earth's microbes died, so would all other life, including the human race,** but if everything else disappeared, microbes would be just fine.

Despite the importance of microbes, scientists have been able to study less than one percent of the estimated millions of microbial species that live on Earth. Mainly, because microbes have strict nutritional requirements and interact with one another in complex ways that currently make it impossible to grow the overwhelming majority of them in a laboratory. Even when we do study them in the field or in our laboratories, they are not life as we know it, they trade genetic material, evolve and mutate differently from anything observable in macroscopic organisms.

There are very practical reasons why it is so important to understand how our existence depends on the microbial world – microbes can be used in disease prevention, industrial processes, environmental remediation, eco-system management and for many other pragmatic purposes.

More important though, is that microbes provide a whole new way to understand the essence and the evolution of life on this planet. Nevertheless, scientists are rapidly advancing our understanding of microbes through the new science of meta-genomics, which involves analyzing DNA content of entire microbial communities rather than analyzing the DNA of individual microbes, as done in laboratory studies. This new science is finally helping scientists explain what microbial communities do and how they live in their natural habitats.

Stages of Bacterial Growth

Under ideal conditions, the growth of a population of bacteria occurs in several stages termed lag, log, stationary, and death. During the lag phase, active metabolic activity occurs involving synthesis of DNA and enzymes, but no growth. Geometric population growth occurs during the log, or exponential phase, when metabolic activity is most intense and cell reproduction exceeds cell death.

Following the log phase, the growth rate slows and the production of new cells equals the rate of cell death. This period, known as the stationary phase, involves the establishment of an equilibrium in population numbers and a slowing of the metabolic activities of individual cells.

The stationary phase reflects a change in growing

condition – for example, a lack of nutrients and/or accumulation of waste products. When the number of cell deaths exceeds the number of new cells formed, the population shifts to a net reduction in numbers and the population enters the death phase, or logarithmic decline phase. The population may diminish until only a few cells remain, or the population may die out entirely.

Temperature and bacteria – the lowest temperature at which a particular species will grow is the minimum growth temperature, while the maximum growth temperature is the highest temperature at which they will grow. The temperature at which growth is optimal is called the optimum growth temperature.

PH and bacteria – like temperature, acid/alkaline balance (pH) also plays a role in determining the ability of bacteria to grow or thrive in particular environments. Most commonly, bacteria grow optimally within a narrow range of pH.

Osmotic pressure and bacteria – osmotic pressure is another limiting factor in the growth of bacteria. Bacteria are 80 – 90 % water and they require moisture to grow because they obtain most of their nutrients from their aqueous environment. Carbon, nitrogen and other growth factors – in addition to water and the correct salt balance – bacteria also require a wide variety of elements, especially carbon, hydrogen, and nitrogen, sulphur and phosphorus, potassium, iron, magnesium and calcium. Growth factors, such as vitamins, pyrimidine's and purines (the building blocks of DNA), are also necessary.

Oxygen – may or may not be a requirement for a particular species of bacteria, depending on the type of metabolism used to extract energy from food (aerobic or anaerobic). In all cases, the initial breakdown of glucose to pyruvic acid occurs during glycolysis, which produces a net gain of two molecules of the energy-rich molecule adenosine triphosphate.

The Ecology of microbes to one another and their surroundings is extraordinary with respect to the diversity of chemical and physical conditions that can be tolerated. Microbes thrive in extreme environments with regards to temperature, high concentrations of salts and sugars, relative acidity, and with or without the presence of oxygen.

Aerobic vs. Anaerobic

There are two distinct types of organisms and tiny single-celled bacteria called aerobic and anaerobic in the human body. Although breathing is essential to life, the specific role that oxygen plays in maintaining life is not easily understood. Basically in organisms that are able to use oxygen, it allows food molecules to be completely broken down, so that every possible bit of energy is extracted.

Organisms that are able to use oxygen for metabolism are called aerobes and since oxygen can actually be rather toxic a cell has to be able to manufacture specific enzymes that detoxify oxygen waste products. Aerobes therefore produce catalase and superoxide dismutase (SOD) for this purpose.

The other types of microbes (bacteria and fungi) are able to live in the absence of oxygen. These organisms either do not have the enzymes required to detoxify oxygen waste or they are not able to make enough of these enzymes to live at normal atmospheric oxygen levels. These microbes, called anaerobes, can still break down food molecules in the absence of oxygen, but cannot do so as efficiently as aerobes.

Anaerobes are able to live in places where aerobes cannot survive, such as the human gut and many other places where oxygen is in low supply. **When pathogenic microbes have access to a normally sterile body fluid or deep tissue that is poorly oxygenated, it can cause serious infection, such as brain abscesses, lung abscesses, aspiration pneumonia, dental infections, etc.**

The identification of anaerobes is highly complex and laboratories may use a different identification system and partial identification is often the goal. For example, there are six species of the Bacteroids genus that may be identified as the Bacteroids fragilis group rather than identified individually.

Organisms are identified by their colonial and microscopic morphology, growth on selective media, oxygen tolerance, and biochemical characteristics.

The differences between aerobic and anaerobic is that aerobic bacteria inhales oxygen to remain alive and anaerobic bacteria may die in the presence of oxygen and therefore avoids oxygen.

To explain a longstanding question on how microbes respond to temperature and oxygen changes that occur when the bacterium enters the gut: it re-acts to change – after sensing it – by switching from aerobic to anaerobic respiration.

For this change to take place, it also experiences a sharp rise in temperature upon entering the host's mouth. This sudden rise in temperature may be the cue for the bacterium to prepare for the subsequent lack of oxygen and upon transition to higher temperature, many of the genes for aerobic respiration were practically turned off and the adaptation to a new environment was accomplished.

Arsenic Eating Bacteria

Scientists examining the toxic waters of a California lake found a new species of bacteria that thrives on poisonous arsenic – a discovery that will surely force some editing of scientific textbooks: The assumption has been that there are six major elements considered essential for life: carbon, hydrogen, nitrogen, oxygen, phosphorus and sulfur. This announcement replaces that theory with a question mark.

If phosphorus is not necessarily needed as a central component of the energy-carrying molecule in all cells, then what about the others? The microbe in question acts very different from life as we knew it in its ability to substitute arsenic for phosphorus.

This dual ability to grow using either element qualifies as a first in the annals of science. Even though this does not qualify as a textbook example of a life from another planet, it will force scientists to re-calibrate their assumptions as they search for other forms of life.

Self-sufficient, invisible, mysterious, deadly – and absolutely essential for all life, they are the Earth's bacteria. No other living thing combines their elegant simplicity with their incredibly complex role: bacteria keep us alive, supply our food, and regulate our biosphere. We cannot live a day without them, and no chemical, antibiotic, or irradiation has successfully eradicated them. We are hopelessly outnumbered by bacteria: they are our partners, even though some of them, under the right conditions, will kill us. No form of life is more important and no form of life is more fascinating.

Chapter 4
MICROBES AT BIRTH

The Journey

It is the journey, not the destination that determines the quality of bacteria a newborn encounters in life's first moments. Infants born via caesarian section had markedly different bacteria on their skin, noses, mouths and rectums, than the ones born vaginally. Those born via C-section miss out on beneficial bacteria passed on by their mothers and may be more likely to develop allergies, asthma and other immune system-related troubles than babies born the traditional way.

Early stages of the body's colonization by microbes shape the developing immune system, help extract nutrients from food and keep harmful bacteria at bay.

Babies born vaginally were colonized predominantly by *Lactobacillus*, microbes that aid in milk digestion. DNA analysis revealed that infants born vaginally carried bacterial populations that matched those of their mother's vaginas, while the C-section babies had a more generic mixture of skin bacteria, similar to that found in hospitals, such as Staphylococcus and Acinetobacter.

First-arrivals to the body are critical for establishing the microbial scene and while the paradigm has been that babies are sterile until birth, recent findings suggest a microbial community already dwelling in the first bowel movement of infants born prematurely.

Estimates of bacterial density during the first week of life revealed no significant link between mode of delivery and onset of colonization, interestingly, however, the babies delivered by caesarian section, before the amniotic membranes had raptured – and thus without exposure to the microbial flora present in their mothers birth canal – had lower bacterial counts than the others for that first week.

Despite individual variations, the profiles for each child showed a surprising degree of continuity – each infant could be recognized by its distinctive microbial flora for weeks and months ahead. By the time babies are one year old, however, their profile converges toward the characteristics adult-like micro-biota, which still retained the stamp of individuality.

The idiosyncratic nature of the early stages of colonization suggests that an infant's initial bacterial profile largely results from incidental microbial encounters, but will likely evolve under strong selection and that certain well-adapted microbes repeatedly win the battle over the opportunistic early colonizers.

The fact is that **at the cellular level we become more microbe than human**, considering that a ninety trillion single cell organisms, representing some 500 bacterial species, inhabit the human gut, **out-populating the cells in our body by a factor of ten**.

It might be of some comfort to know that in exchange for room and board, these microbes offer several essential services, from pathogen protection to nutrient metabolism, and likely others yet to be discovered.

Given the importance of beneficial bacteria for a healthy baby it is essential that the mother's health is ensured, preferably before pregnancy. A healthy mother is more likely to deliver a healthy baby.

Newborns and Breast Milk

Doctors have long known that infants who are breast-fed contract fewer infections than those who are given formula. Humans are born extremely immature, with the major organs and immune system not fully developed.

For its survival the infant depends on an extraordinary well-adapted evolutionary strategy shared by all mammals, breast-feeding. But what does milk contain that makes it so essential for the newborn and how does it provide immunity, nutrition, and a source for optimal growth?

Human milk is a very complex living fluid which comprises proteins, carbohydrates, lipids, cells and other biologically important components. These milk components interact synergistically with each other and their environment (the infants gut) at a bio-molecular level with the final result being that breast-milk feeds and protects the newborn.

Human milk may reduce the incidence of disease in infancy because mammalian evolution promotes a survival advantage. In addition factors in milk promote gastrointestinal mucosal maturation, decrease in incidence of infection, alter gut microflora, and have immune modulatory and anti-inflammatory functions. Hormones, growth factors and cytokines in human milk may modulate the development of disease. Furthermore breast-fed infants have reduced exposure to foreign dietary antigen.

Following the termination of breast-feeding there is evidence of ongoing protection against illness due to protective influences on the immune system mediated via human milk. Industry continues to attempt to improve infant formula with the addition of compounds such as fatty acids, oligosaccharides, nucleotides and lactoferrin.

However, human milk has such far-reaching effects on the infant's immune response that optimal development depends heavily on its provision. Most physicians presumed that breast-fed children fared better simply because milk supplied directly from the breast is free of bacteria. Yet even infants who receive sterilized formula suffer from more meningitis and infection in the gut, ear, respiratory tract and urinary tract.

All newborns receive some coverage in advance of birth. During pregnancy, the mother passes antibiotics to her fetus through the placenta. These proteins circulate in the infant's blood for weeks to months after birth, neutralizing microbes or marking them for destruction by phagocytes-immune cells that consume and break down bacteria, viruses and cellular debris.

Because the mother makes antibodies to pathogens in her environment, the baby receives the protection it most needs against infectious agents it is most likely to encounter in the first few weeks of life.

Anti-bodies delivered to the infant ignore useful bacteria normally found in the gut. This flora serves to crowd out the growth of harmful organisms, thus providing another measure of resistance. All things considered, breast-milk is truly a fascinating fluid that supplies infants with far more than nutrition.

It protects against infection through specific and nonspecific immune factors and has long-term consequences for metabolism and disease later in life. Human milk enhances the immature immunologic system of the neonate and strengthens host defense mechanisms against infective and other foreign factors in **human milk. It protects the infants against infection until they can protect themselves.**

Once other foods than breast-milk are introduced, the established microbes are ready to adapt and utilize the nutrients from a new source and should be able to maintain a healthy balance. **Milk microbes may be the most important factor to our nutrition and well-being.**

Chapter 5
NORMAL HUMAN FLORA

Organisms Regularly Found

The mixture of organisms regularly found at any anatomical site is referred to as the normal flora, except by researchers in the field who prefer the term "indigenous micro-biota". The normal flora of humans consists of a few eukaryotic fungi and protists, but bacteria are the most numerous and obvious microbial components.

Populations of microbes (such as bacteria and yeasts) inhibit the skin and mucosa (mucous membranes) and they are the body's first lines of defense against illness and injury.

Their health depends upon the delicate balance between our own cells and the millions of bacteria and other one-celled microbes. Their role forms part of normal, healthy human physiology, however if microbe numbers grow beyond their typical ranges (often due to a compromised immune system) or if microbes populate atypical areas of the body (such as through poor hygiene or injury), disease can result.

Bacterial cells are much smaller than human cells, and with at least ten times as many bacteria as human cells in the body, members of the flora are found on all surfaces exposed to the environment (on the skin and eyes, in the mouth, nose and small intestine), the vast majority of bacteria live in the large intestine. It is estimated that 500 to 1000 species of bacteria live in the human gut and roughly a similar number on the skin.

E. coli is the best known bacterium that regularly associates itself with humans, being an invariable component of the human intestinal tract. Even though E. coli is the most studied of all bacteria, and we know the exact location and sequence of 4,288 genes on its chromosome, we do not fully understand its ecological relationship with humans.

Many of the bacteria in the digestive tract, collectively referred to as the gut flora, are able to break down certain nutrients such as carbohydrates that humans otherwise could not digest. The majority of the commensal bacteria are anaerobes, meaning they survive in an environment with no oxygen. In fact, not much is known about the nature of the associations between humans and their normal flora, but they are thought to be dynamic interactions rather than associations of mutual indifference.

Both host and bacteria are thought to derive benefits from each other, and the associations are, for the most part, mutualistic. The normal flora derive from their host a steady supply of nutrients, a stable environment, protection and transport. The host obtains from the normal flora certain nutritional and digestive benefits, stimulation of the development and activity of immune system, and protection against colonization and infection by pathogenic microbes.

While most of the activities of the normal flora benefit their host, some of the normal flora are parasitic (live at the expense of their host), and some are pathogenic (capable of producing disease). Diseases that are produced by the normal flora in their host may be called endogenous diseases. Most endogenous bacterial diseases are opportunistic infections, meaning that the organism must be given a special opportunity of weakness or let-down in the host defenses in order to infect. An example of an opportunistic infection is chronic bronchitis in smokers wherein normal flora bacteria are able to invade the weakened lung.

Sometimes the relationship between a member of the normal flora and its host cannot be deciphered. Such a relationship where there is no apparent benefit or harm to either organism during their association is referred to as a commensal relationship. Many of the normal flora that are not predominant in their habitat, even though always present in low numbers, are thought of as commensal bacteria. However, if a presumed commensal relationship is studied in detail, parasitic or mutualistic characteristics often emerge.

Tissue specificity – most members of the normal bacterial flora prefer to colonize certain tissues and not others. This "tissue specificity" is usually due to properties of both the host and the bacterium. Usually, specific bacteria colonize specific tissues by one or another of these mechanisms.

Tissue tropism – is the bacterial preference or predilection for certain tissues for growth. One explanation for tissue tropism is that the host provides essential nutrients and growth factors for the bacterium, in addition to suitable oxygen, pH, and temperature for growth.

Specific adherence – most bacteria can colonize a specific tissue or site because they can adhere to that tissue or site in a specific manner that involves complementary chemical interactions between the two surfaces and bio-chemical interactions between bacterial surface components (ligands or adhesins) and host cell molecular receptors. The bacterial components that provide adhesins are molecular parts of their capsules, fimbriate, or cell walls. The receptors on human cells or tissues are usually glycoprotein molecules located on the host cell or tissue surface.

Bio-film formation – some of the indigenous bacteria are able to construct bio-films on a tissue surface, or they are able to colonize a bio-film built by another bacterial species. Many bio-films are a mixture of microbes, although one member is responsible for maintaining the bio-film and may predominate. The classic bio-film that involves components of the normal flora of the oral cavity is the formation of dental plaque on the teeth. Plaque is a naturally-constructed bio-film, in which the consortia of bacteria may reach a thickness of 300-500 cells on the surfaces of the teeth. These accumulations subject the teeth and gingival tissues to high concentrations of bacterial metabolites, which result in dental disease.

The composition of the Normal flora in Humans is exceedingly complex and its makeup may be influenced by various factors, including genetics, age, sex, stress, nutrition and diet of the individual.

Three developmental changes in humans, weaning, the eruption of teeth, and the onset and cessation of ovarian functions, invariably affect the composition of the normal flora in the intestinal tract, the oral cavity and the vagina, respectively. However the limits of these fluctuations, the bacterial flora of humans is sufficiently constant to a given general description of the situation.

Normal Flora of the Skin

It has been calculated that the surface of a human adult is covered with approximately 2 square meters of skin. The density and composition of the normal flora of the skin varies with anatomical locale. The high moisture content of the axilla, groin, and areas between the toes supports the activity and growth of relatively high densities of bacterial cells, but the density of bacterial populations at most other sites is fairly low, generally in the hundreds or thousands per square cm. The greatest influence on bacterial populations appears to be body location. For example, the bacteria under our arms are likely more similar to those of another person's underarm than they are to the bacteria on our own forearm.

Most bacteria on the skin are sequestered in sweat glands. Skin microbes found in the most superficial layers of the epidermis and the upper parts of the hair follicles (Staphylococcus epidermids, Micrococcus sp. and Propionibacterium sp.) are generally non-pathogenic and considered to be commensal, although mutualistic and parasitic roles have been assigned to them. For example staphylococci and propionic bacterium produce fatty acids that inhibit the growth of fungi and yeast on the skin. But, if Propionibacterium acnes, a normal inhabitant of the skin, becomes trapped in hair follicles, it may grow rapidly and cause inflammation and acne.

Normal Flora of the Conjunctiva (Eye)

A variety of bacteria may be cultivated from the normal conjunctiva, but the number of organisms is usually small. The conjunctiva is kept moist and healthy by the continuous secretion from the lachrymal glands. Blinking wipes the conjunctiva every few seconds mechanically washing away foreign objects, including bacteria. Lachrymal secretion (tears) also contain bactericidal substances including lysozyme. There is little or no opportunity for microorganisms to colonize the conjunctiva without special mechanisms to attach to the epithelial surfaces and some ability to withstand attack by lysozyme. Pathogens which do infect the conjunctiva are thought to be able to specifically attach to the conjunctival epithelium.

Newborn infants may be especially prone to bacterial attachment, since Chlamydia and Neisseria might be present on the cervical and vaginal epithelium of an infected mother, silver nitrate or an antibiotic may be put into the newborn's eyes to avoid infection after passage through the birth canal.

Normal Flora of the Respiratory Tract

A large number of bacterial species colonize the upper respiratory tract (nasopharynx). The nares (nostrils) are always heavily colonized, but the healthy sinuses are sterile. The lower respiratory tract is virtually free from microorganisms, mainly because of the efficient cleansing action of the ciliated epithelium which lines the tract.

Any bacteria reaching the lower respiratory tract are swept upward by the action of the mucociliary blanket that lines the bronchi, to be removed subsequently by coughing, sneezing, swallowing, etc. If the respiratory tract epithelium becomes damaged, as in bronchitis or viral pneumonia, the individual may become susceptible to infection by pathogens descending from the nasopharynx.

Normal Flora of the Urogenital Tract

Urine is normally sterile, and since the urinary tract is flushed with urine every few hours, microorganisms have problems gaining access and becoming established. The flora of the anterior urethra, as indicated principally by urine cultures, suggests that the area may be inhabited by a relatively consistent normal flora.

Their numbers are not plentiful, however, in addition some enteric bacteria, which are contaminants from the skin, vulva or rectum, may occasionally be found at the anterior urethra. The vagina becomes colonized soon after birth and during reproductive life, from puberty to menopause, the vaginal epithelium contain glycogen due to the actions of circulating estrogen. Doderlein's bacillus predominates, being able to metabolize the glycogen to lactic acid and the resulting low pH of the vaginal epithelium prevents the establishment of most other bacteria as well as the potentially pathogenic yeast, Candida albicans. This is a striking example of the protective effect of the normal bacterial flora for their human host.

Normal Flora of the Oral Cavity

The presence of nutrients, epithelial debris, and secretions makes the mouth a favorable habitat for a great variety of bacteria and presents a succession of different ecological situations with age. At birth, the oral cavity is composed solely of the soft tissue of the lips, cheeks, tongue and palate, which are kept moist by the secretions of the salivary glands.

The eruption of teeth during the first year leads to S.mutans and S.sanguis, which require a non-epithelial surface in order to colonize and will persist as long as teeth remain. The creation of the gingival crevice area (supporting structure of the teeth) increases the habitat for the variety of anaerobic species found. The complexity of the oral flora continues to increase with time, and bacteroides and spirochetes colonize around puberty.

The normal bacterial flora of the oral cavity clearly benefit from their host who provides nutrients and habitat. There may be benefits, as well, to the host. The normal flora occupy available colonization sites which makes it more difficult for other microorganisms to become established and contribute to host nutrition through the synthesis of vitamins and to immunity by introducing low levels of circulating and secretory antibiotics that may cross react with pathogens.

Normal Flora in the Gastrointestinal Tract

The bacterial flora of the gastrointestinal (GI) tract in humans show differences in the composition which are influenced by age, diet, cultural conditions, and the use of antibiotics. The latter greatly perturbs the composition of the intestinal flora. In the upper GI tract the esophagus contains only the bacteria swallowed with saliva and food. Because of the acidity of the gastric juice, very few bacteria can be cultured from the normal stomach.

However, half of the population in North America may be colonized by a bacterium, *Helicobacter, pylori.* Since the 1980s, this bacterium has been known to cause gastric ulcers, and is probably a cause of gastric and duodenal cancer as well. The Australian microbiologist, Barry Marshall, received the Nobel Prize in Physiology and Medicine in 2005 for demonstrating the relationship between *Helicobacter* and gastric ulcer.

The proximal small intestine has a relatively sparse flora, consisting mainly of lactobacilli and faecalis, while the distal part of the small intestines contains greater numbers of bacteria, including coliforms (*E.coli and relatives*) and *Bacteroids*, in addition to lactobacilli and enterococci.

The flora of the large intestine (colon) is quantitatively similar to that found in feces and is our only direct association with archaea as normal flora. The composition of the flora of the gastrointestinal tract varies along the tract (at longitudinal levels) and across the tract (at horizontal levels) where certain bacteria attach to the epithelium and others occur in the lumen.

There is frequently a very close association between specific bacteria in the intestinal ecosystem and specific gut issues or cells. **The gut flora is the human flora of micro-organisms that normally live in the digestive tract and can perform a number of useful functions for the host**. The metabolic activity performed by these bacteria is equal to that of a virtual organ, leading to gut bacteria being termed a forgotten organ.

It is in the intestinal tract that we see the greatest effect of the bacterial flora on their host. This is due to their large mass and numbers. Bacteria in the human GI tract have been shown to produce vitamins and may otherwise contribute to nutrition and digestion. But their most important effects are in their ability to protect their host from establishment and infection by alien microbes and their ability to stimulate the development and the activity of the immunological tissues.

On the other hand, some of the bacteria in the colon (*e.g. Bacteroids*) have been shown to produce metabolites that are carcinogenic, and there may be an increased incidence of colon cancer associated with these bacteria. Alterations in the GI flora brought on by poor nutrition or perturbance with antibiotics can cause shifts in populations and colonization by non-residents that leads to gastrointestinal disease.

Chapter 6
THE DENTAL CONNECTION

Dental Plaque, Caries and Periodontal Disease

The most frequent and economically-important condition in humans resulting from interactions with our normal flora is probably dental caries. Dental plaque, dental caries, gingivitis and periodontal disease result from actions initiated and carried out by the normal bacterial flora.

Dental Plaque – which is material adhering to the teeth, consists of bacterial cells (60-70% the volume of plaque), salivary polymers, and bacterial extracellular products. Plaque is a naturally constructed biofilm, in which the consortia of bacteria may reach a thickness of 300-500 cells on the surface of the teeth. These accumulations subject the teeth and gingival tissues to high bacterial metabolites, which result in dental disease.

Dental Caries – is the destruction of the enamel, dentin or cementum of teeth, due to bacterial activities. Caries are initiated by direct demineralization of the enamel of teeth due to lactic acid and other organic acids which accumulate in dental plaque. Lactic acid bacteria in the plaque produce lactic acid from the fermentation of sugars and other carbohydrates in the diet of the host. These organisms normally colonize the occlusal fissures and contact points between the teeth, and this correlates with the incidence of decay on these surfaces. After initial weakening of the enamel, various bacteria gain access to the interior regions of the tooth.

Periodontal Diseases – are bacterial infections that affect the supporting structures of the teeth (gingiva, cementum, periodontal membrane and alveolar bone). The most common form, gingivitis, is an inflammatory condition of the gums. It is associated with accumulations of bacterial plaque in the area. Increased population of Actinomyces have been found, and they have been suggested as the cause. Diseases that are confined to the gum usually do not lead to loss of teeth, but there are other more serious forms of periodontal disease that affect periodontal membrane and alveolar bone resulting in bone loss.

Bacteria in these lesions are very complex populations consisting of Gram-positive organisms (including Actinomyces and Streptococci) and Gram-negative organisms (including Spirochetes and Bacteroides). The mechanisms of tissue destruction in periodontal disease are not clearly defined but hydrolytic enzymes, endotoxins, and other toxic material metabolites seem to be involved.

Acupuncture Meridians and Dentistry

Conventional dental procedures offer a technique which does not take into account bio-compatibility of the filling materials, potential injury to the surrounding tissues due to the caustic nature of medicaments used and high percentage of residual bacterial contamination. **According to research 75% of root canaled teeth have residual bacteria remaining in the dentinal tubules.**

These lingering infections produce toxic wastes that enter the blood stream and can affect any part of the body. Root canals are part of dentistry called endodontic, which is concerned with the pathology of dental pulp and the area surrounding the root. A root canal is a procedure to allow a tooth that is painful or no longer viable because of nerve damage or death to remain in the mouth.

Most dentists consider root canals an advance in dentistry, a superior alternative to removal of a seriously compromised tooth. However a growing number of physicians, including dentists, believe that root canals can be the cause of, or at least contribute to a long list of illnesses, including cancer.

The fallacy with the concept of keeping a dead tooth in the mouth is that the body does not like dead things in it and will try, sometimes desperately, to get rid of it. Notwithstanding the fact that it may be a tooth it does not change the fact that a root canal can devastate the human immune system. This is due to the interference of the flow of bio-electrical energy through acupuncture meridians associated with all the organs of the body.

In the 1950's, Reinhold Voll, a German M.D. using an electro-acupuncture biofeedback system he had developed, discovered that each tooth in the mouth relates to a specific acupuncture meridian. **He found that if a tooth became infected or diseased, the organ**

on the meridian would also become unhealthy.

Conversely – he found that a diseased organ could cause a problem with the corresponding tooth. With a root canal, or even a filling, a crown or anything that is not compatible with the body, it sets up an interference field, blocking the energy flow (chi) passing through this meridian and causes a disease in an organ or body function remote to the tooth. For example a front upper incisor is on the Kidney / Bladder meridian and having a root treated tooth here may cause gynecological problems, kidney problems, impotence, and sterility if you follow a Chinese Medicine theme.

These teeth also relate to spinal segments and joints, the front incisor relates to the coccyx and posterior knee and to L2, L3, S3, and S6 in the spine.

If the tooth is removed, the energy does tend to pass through it, however, without the stimulation from the tooth, blood circulation and lymphatic drainage will be impaired, and the bone and tissue surrounding the extraction site can become diseased and die.

Infections of the teeth and toxins, have no place to go but down, down into the jaw bone and into the rest of the body, creating systemic pathology. **What is interesting about this is, that breast cancer patients had root canals or other infections, on the same acupuncture meridian**. Since the microbes do not originate in the root canal, what this implies is that breast cancer cannot form without the assistance of the metals and microbes coming from the root canals.

Of equal interest is the relationship of root filled

teeth to Traditional Chinese Medicine and body energies. It is assumed in dentistry that the extent of bone loss is a direct indication of the amount of infection present. This is a false assumption because the bone loss may take time to develop. The extent of the bone loss about the end of the root is also a function of the body's immune system being able to isolate the infection process. It has little to do with the degree of infection.

Sometimes there is no bone loss, but instead, a condensation of bone about the end of the tooth. Dentists are taught that this indicates a lack of infection. The reality is that teeth showing a Condensing Osteitis are demonstrating that the body's immune system is incapable of quarantining the infection locally. **These are often the teeth which cause the greatest systemic effects.**

Due to the caustic nature of substances used and the high percentage of residual bacterial infections remaining in the dentinal tubules there is no drug, homeopathic remedy, vitamin or mineral that can effectively kill these tiny bacteria. Only the use of bio-frequencies has the capacity of penetrating the surrounding bone and root without damage to tissues. Even then there is no way to stop new bacteria from entering these tubules from the oral cavity again. **These lingering infections produce the most toxic substances known to biochemistry and toxicology that enter the blood stream and can affect every part of the body.**

A tooth is an organ, just as any other organ or bone in the body. **An abscessed or gangrenous tooth**

ıs not only a dead tooth, it is a dead organ.

The problem arises because these teeth are dead and prone to infection and can threaten to infect surrounding tissue, including the jawbone, possibly triggering cavitation's. Today we know that the toxins made by the bacteria that live in the billions in root canal teeth contain the most toxic organic substance known to man, thio-ethers. **Thio-ethers are 1000 times more toxic than botulism toxin, which used to be considered the most toxic organic substance.**

One would be than well advised to worry less about anthrax and smallpox, and instead focus on root canals which are much more likely to cause personal harm. In addition to thio-ethers, other severe toxins from these root-canal bacteria include thio-ethanol's and mercaptans which have been found in the tumors of women who have breast cancers, draining through the lymphatic system down the cervical chain of lymph nodes and ultimately in to the breast tissue.

Studies on thousands of teeth have demonstrated the presence of bacteria in 80 to 90% of the root canals after they have been sterilized. The problem is that root canal therapy cannot sterilize the inside of the tooth. **As a result, the trapped bacteria mutate and migrate to infect the heart, kidneys, eyes, stomach, and countless other body tissues**.

Since there are varying degrees of oxygen deprivation, each level of oxygen stimulates the bacteria to mutate into a slightly different bacterium, a whole plethora of these can develop from just one bacterium. All but five are classified as anaerobic, or those that thrive in the absence of oxygen.

These bacteria produce toxic waste products, which can be picked up by the drainage system at the apex of the tooth, or flow down the dentin tubules into the periodontal ligament, from there into the fluids around the tooth and then into the bloodstream. Besides being harbored in root canals, these dangerous bacteria also take up residence in cavitation's which result from most extracted teeth.

The energetic relationship between teeth and the rest of the body is opening whole new avenues of dental care and the necessity to incorporate other complimentary therapies.

Acupuncture and the Immune System

Our immune system is responsible for keeping bacteria, viruses, microbes and toxins from interrupting the normal functions of the body. It is a complex system that involves specialized cells (such as T-cells, B-cells, macrophages and others) that recognize and destroy foreign entities.

Poor nutrition, stress, lack of sleep and repeated exposure to harmful agents all can lead to a declining immune system that is no longer able to respond effectively when the need arises. And once it is compromised, a weak immune system results in frequent colds, allergies, the inability to fight off infections and general feelings of malaise.

Acupuncture and Chinese herbal therapy are effective tools to strengthen a weakened immune system and have far-reaching effects from increasing red and white cell counts in compromised individuals to increasing lymphocyte proliferation and natural killer cell activity. This is accomplished through the needling or stimulation of specific acupuncture points and their subsequent effect on the nervous system, tailored to the individual to produce best results to support and strengthen a healthy immune system.

Chapter 7
MICROBIOLOGY

Early Observations

Throughout most of history, people had no knowledge of the minute organisms that exist everywhere, and the implications of this ignorance were huge. Infectious diseases could not be understood, prevented or cured, prior to the discovery of microorganisms.

Modern infectious disease medicine is now built upon a solid understanding of microbiology with its long, rich history, initially centered in the cause of infectious diseases but now including practical applications of the science. Many individuals have made significant contributions to the development of microbiology.

Historians are unsure who made the first observation of microorganisms, but the microscope was first developed during the mid-1600s, when an English scientist named Robert Hooke made key observations. He is reputed to have observed strands of fungi among the specimens of cells he viewed. In the 1670's and the decades thereafter, Anton van Leeuwenhoek, a Dutch merchant and science hobbyist, constructed some of the first high powered microscopes and made careful observations of microscopic organisms, which he called animalcules. His excellent scopes revealed a world of life too small to be seen by the naked eye.

Leeuwenhoek's discoveries were the beginning of the end for spontaneous generation, the belief that life could magically spring from non-living matter on a daily basis. His findings inspired other scientists to investigate how tiny living things seem to appear from nowhere – in rainwater, spoiled food and the human body – and led to the understanding that life springs from life.

Until his death in 1723, van Leeuwenhoek revealed the microscopic worlds to scientists of the day and is regarded as one of the first to provide accurate descriptions of protozoa, fungi, and bacteria.

Germ Theory

This explosion of scientific inquiry into the microscopic world resulted in a dramatic change in medicine called Germ Theory, the knowledge that some diseases were caused by microbes.

At the most basic level, all living things can be divided into two major groups of organisms, depending on the type of cells they have. Those organisms which contain more complex cells and which have a nuclei are called eukaryotes (true kernel or nucleus). These organisms include animals, plants, fungi and protists. The other group of organisms are less complex, lack nuclei and are called prokaryotes. The prokaryotes are also known as bacteria or as monerans.

Organisms that are typically studied by microbiologists fall into a few basic categories: fungi, protozoa, algae, bacteria, archaea and viruses.

Fungi: These eukaryotic organisms can be quite large, like the underground fungi that generate mushrooms, or they can be very small, such as tiny molds of yeasts. As important de-composers of dead organisms, fungi are essential for recycling matter in living systems, but there are also fungi that cause disease.

Protozoa: Protozoans are single-celled eukaryotes that have animal-like characteristics in their cell structure as well as how they procure food and move about. Members of this group include amoebae, flagellates, ciliates, and sporozoans. Most protozoans live in water, but some live in human hosts and are able to cause disease.

Algae: These water-loving eukaryotes can range in size from microscopic, single-celled organisms to giant kelp. Algae are more plant-like than animal-like, in that they make their own food from carbon and sunlight energy (photosynthesize). Most algae are beneficial, even critical for life on earth. Few cause disease.

Bacteria: Bacteria are single-celled prokaryotic organisms. They have cell walls made of peptidoglycan, a molecule not found anywhere else in nature. Bacteria are all around us, some are normal flora that naturally inhibit the human body and are helpful or at least not harmful and are also utilized to make enzymes for detergent and antibiotics against other types of bacteria. But there are also many bacterial pathogens that cause disease.

Archaea: This group of prokaryotic microbes is very similar to bacteria. However their cell walls are not made of peptidoglycan, and there are no known archaea that cause disease in humans.

Viruses: Viruses are a-cellular, non-living infectious agents. They are smaller than cells and are strictly parasitic. Viruses cannot reproduce on their own and they must first find a cell to infect, which will then function as a factory for making more viruses. Viruses are so small that they cannot be seen with a compound microscope, the type of scopes typically used in laboratories to see small organisms, these minute microbes can only be seen using a high-tech electron microscope.

History of Microbiology

Louis Pasteur (1822-1895) performed numerous experiments to discover why wine and dairy products became sour, and he found that bacteria were to blame. Pasteur called attention to the importance of micro-organisms in everyday life and stirred scientists to think that if bacteria could make wine sick, then perhaps they could cause human illness. His work also encouraged the belief that micro-organisms were in the air and could cause disease.

Pasteur's attempts to prove the germ theory were unsuccessful. However, the German scientist Robert Koch provided the proof by cultivating anthrax bacteria apart from any other type of organism. He then injected pure culture of bacilli into mice and showed that the bacilli invariably cause anthrax. The procedures used by Koch came to be known as Koch's postulates. They provided a set of principles whereby other micro-organisms could be related to other diseases.

In the late 1800's and for the first decade of the 1900s, scientists seized the opportunity to further develop the germ theory of disease as enunciated by Pasteur and proved by Koch. There emerged a Golden Age of Microbiology during which many agents of different infectious diseases were identified. Many of the etiologic agents of microbial disease were discovered during that period, leading to the ability to halt epidemics by interrupting the spread of micro-organisms.

Despite the advances in microbiology, it was rarely possible to render life-saving therapy to an infected patient. Then, after World War II, the antibiotics were introduced to medicine. The incidence of pneumonia, tuberculosis, meningitis, syphilis, and many other diseases declined with the use of antibiotics.

In the 1940s, the electron microscope was developed, cultivation methods for viruses were introduced and the knowledge of viruses expanded rapidly. With the availability of vaccines in the 1950s and 1960s, such viral diseases as polio, measles, mumps, and rubella came under control.

Modern microbiology reaches into many fields of human endeavor, including the development of pharmaceutical products, the use of quality-control methods in food and dairy production, the control of disease-causing micro-organisms in consumable waters, and the industrial applications of microbes.

Micro-organisms are used to produce vitamins, amino acids, enzymes, and growth supplements. They manufacture many foods, including fermented dairy products (sour cream, yogurt, and buttermilk), as well as foods such as pickles, sauerkraut, breads, and alcoholic beverages.

One of the major areas of applied microbiology is biotechnology. In this discipline, micro-organisms are used as living factories to produce pharmaceuticals that otherwise could not be manufactured. **These substances include the human hormone insulin, the antiviral substance interferon, numerous blood-clotting factors and clot-dissolving enzymes, and a number of vaccines.** Bacteria can be re-engineered to increase plant resistance to insects and frost, and biotechnology will represent a major application of micro-organisms in the future.

A recent discovery has shown that bacteria can also be divided into two distinct groups, of which one, the archea bacteria is more closely related to humans, than the other group, the eubacteria. With the advent of molecular genetics and recombinant DNA technology, Bacteria now play a very important role as producers of human substances.

Since we have learned how genes function, we are able to introduce a human gene into bacterium and have the product of the human gene expressed. Consequently, a hormone called erythropoietin, which is absolutely necessary for the proper development of red blood cells (erythrocytes), but very difficult to isolate, is now available in high quantity.

People who do not have kidneys, cannot make

this hormone, however, because it has been cloned into bacteria, plenty of it can be made, purified, and given to these individuals. Human insulin can be made similarly. These are only two examples of the many substances now available to treat human disorders because of our understanding of bacteria.

DNA – the Genetic Blue Print

At the center of every living creature on the planet is a blueprint of its genetic makeup. This is called DNA and it influences everything from the organisms' physical structure to its appearance and even its mental processes. It is interesting to note however, that for all our vast differences, our DNA is actually a lot more alike than different. However it is those slight disparities that make humans so different from animals.

Given the vast physical differences between humans and animals, it is interesting to note just how similar DNA is, a code comprised of 4 nucleotides or types of molecules known as adenine, cytosine, guanine and thymine. The cellular machinery in the body reads these molecules and creates a backup of the RNA, which is translated into more complex molecules called amino acids, which are essential components to protein formation.

There is basically no difference between human and animal DNA except for its underlying codes and the actual amount of DNA present, which may vary from species to species and the order in which the code is assembled.

In each and every single creature, the order of the

previously mentioned adenine, guanine, cytosine, and thymine are switched around and this order will determine what the organism will be, and how it will work. Depending on this arrangement, the organism may develop scales, appendages, skin, and various other body parts and features.

The genomes or DNA of microbes contain defined DNA patterns called genome signatures. Such signatures may be used to establish relationships and to search for DNA from viruses or other organisms in the microbe's genomes. Foreign DNA bacteria has often been associated with disease-causing abilities.

The increasing number of sequenced genomes that has become available during recent years has made it possible to give us more reliable information on how different microbes DNA composition is influenced by environment and lifestyle. This information can also be used to deepen our understanding of the evolutionary development that follows natural selection at the DNA level.

Such knowledge is absolutely necessary to understand mechanisms leading to bacteria becoming pathogenic and resistant to antibiotics. The spread of antibacterial resistance often occurs through vertical transmission of mutations during growth and by genetic recombination of DNA by horizontal exchange.

Bacteria and Gene Expression

According to one analysis, 463 human genes are changed during an infection with Mycobacterium tuberculosis. Of the 463 genes whose expression were changed, 366 of them were known genes, the other were unknown mutations. It is quite possible that auto-immunity in which it is believed that the body is attacking itself, is caused by bacterial-induced alteration of human genes.

All the bacterium would need to do in order to generate an apparent autoimmune reaction would be to interfere with the genes necessary for the production of proteins against which auto-antibodies are produced. Pathogenic bacteria have a variety of ways disrupting the activity of and causing damage to human genes.

Horizontal gene transfer – Bacteria can insert DNA into human DNA

Interruption of transcription – Translation of DNA and RNA

Disruption of DNA – Disruption of the DNA repair mechanism and translation of DNA.

Given the rapid diversification in the microbial world, it has become increasingly difficult to classify bacteria with traditional approaches. When it comes to bacteria, the very definition of species may have to be reconsidered.

Microbes are extremely diverse and represent all the great kingdoms of life, including animals, plants, fungi, protists and bacteria. In fact, in terms of numbers, most of the diversity of life on earth is represented by microbes, the oldest organisms on earth and among the estimated three billion species, only about 5% have been discovered. **Without microbes, existence of all other living organisms would be impossible.**

Chapter 8
BIOLOGICAL TERRAIN

The Theory of Disease

The germ theory of disease as popularized by Louis Pasteur, the father of modern germ theory and the inventor of pasteurization says that there are fixed germs (or microbes) which invade the body and cause a variety of separate, definable diseases.

In order to get well, we need to identify and then kill whatever germ caused the sickness. The tools generally employed are pharmaceuticals, surgery, radiation and chemo-therapy. Prevention include the use of vaccines as well as drugs, which theoretically at least – work by keeping germs at bay.

At about the same time as Pasteur was promoting his single form (mono-morphic) germ theory, a contemporary be the name of Claude Bernard (1813-1878) was developing the theory that the body's ability to heal was dependent on its general condition or internal environment. Thus disease occurred only when the terrain or internal environment of the body became favorable to germs.

Another extremely brilliant contemporary of Claude Bernard's was Antoine Bechamp who built upon and extended Bernard's idea, developing his own theory of health and disease which revolved around the concept of pleo-morphism.

In contrast to Pasteur's mono-morphic or single-formed, fixed state microbes, Bechamp had discovered tiny organisms which were pleo-morphic or many-formed (pleo = many and morph = form). Interestingly these microbes were found to be present in all things whether living or dead, and they persist even when the host has died. Many were impervious to heat as well.

These microbes appear also when and if the terrain is disturbed from hemostatic equilibrium. This bio-terrain is the concept of pleo-morphism. The archetypal recognition of meta-morphosis is deeply embedded in the human psyche, and most likely takes root in the nature of being and becoming. A fertilized ovum changes from embryo to fetus into a full human form. An acorn becomes an oak tree, a cloud transforms into rain and summer into winter. The cyclic mutagenic essence of things is encoded in myth, music and art, and above all, perhaps in the Chinese classic book of changes, the I Ching.

This primary logic of change and mutation is the basis for the theory of pleo-morphism, as introduced by Antoine Bechamp and later developed by Gunther Enderlein in Germany and Gaston Naessens in Canada. Instead of a micro-organism existing in a single, unchanging morphological state, microbes are understood to undergo radical morphological changes conditioned by alterations or fluctuations in the bio-terrain.

Then it becomes of utmost therapeutic importance,

not so much to kill the microbe (the anti-biotic tactic used by allopathic physicians, and further extended to attack viruses, yeasts, fungi and cancer cells), but rather to adjust the internal milieu back towards a healthier state. If this is done, the virulent forms of the microbe will automatically begin to regress back towards their state of origin.

It is ironic, that conventional bio-medicine, which denies the reality of pleo-morphism and prefers the illusion of a fixed-target enemy – the identifiable and static germ – has ended up generating a cascade of dangerous mutations in microbes as a result of its targeted approach. It is well known and a worrying fact that today, increasing numbers of bacterial diseases, once treatable with antibiotics, have recently developed drug resistance, some of these to the point where antibiotics now fail to make any impact. The same is true for viruses.

Aggressive anti-microbial treatments with harsh pharmaceutical cocktails have only succeeded in triggering the emergence of super bugs, whether bacterial, fungal or viral.

Our bodies are in effect mini eco-systems, or biological terrains in which nutritional status, level of toxicity and pH or acid/alkaline balance play key roles. For this and other reasons Bechamp argued strenuously against vaccines, asserting that the most serious disorders may be provoked by the injection of living organisms into the blood.

Decades after Pasteur's death the germ theory is believed to be the central cause of disease because around it exists a colossal supportive infrastructure of commercial interests. Interestingly to this day, the whole theory of microbes and how they operate has never been disproved – or proven to be false – by opposing research.

To the contrary, decades of research – beginning with Pasteur himself – has only served to bolster the germ theory. Not only does the germ theory remain unsubstantiated today, but Louis Pasteur himself, recanted on his death bed "**The microbe is nothing, the terrain is everything** ".

Regeneration of the bio-terrain – residential bacteria perform the function of helping with assimilation and metabolic synthesis of vital nutrients and by occupying niches, they keep potentially harmful micro-organisms at bay.

They even work synergistically with lymphocytes (white blood cells) in mysterious alliances to take out parasitic infections and clear the body of toxins. The notion that the interior of the human body is intrinsically sterile flies in the face of overwhelming evidence, though it is prevailing dogma to allopathic medicine to attempt to kill off microbes. Yet these organisms, the oldest form of biological existence on the planet, undoubtedly have accumulated vast reservoirs of intelligence and adaptive understanding.

All life-sustaining information needed by cells is supplied via the extracellular fluid to the appropriate cell membrane receptors. The extracellular fluid contains ionized water, supporting cellular metabolism and polarity. Similar to the human life form enclosed in skin, the cell is enclosed in a membrane, with a constant temperature, bathed in fluid similar to the original content of seawater. Our internal biology depends upon ionized mineral salts for physiological functions, restoring minerals and water balance is key to establishing the balance of the internal environment – the biological terrain.

If our terrain is out of balance due to an unhealthy lifestyle or xenobiotics (environmental toxins), we are setting ourselves up for disease. It is essential to maintain a healthy balance of beneficial micro-organism.

According to Hulda Clark, 2006, no matter how long and confusing the list of symptoms is, from chronic fatigue to infertility to mental problems, sickness is caused by either toxins or parasites.

Probiotic Supplements

There are a few different terms that are used when discussing probiotics and their beneficial properties.

Prebiotic – (before life) a substance that cannot be digested, but promotes the growth of beneficial bacteria or probiotics.

Probiotic – (for life) a substance that contains micro-organisms or bacteria that are beneficial to the host organism.

Symbiotic – (plus life) a substance containing both prebiotic and probiotic.

It may seem strangely related but the benefits of probiotics have been extended to cognitive and emotional health.

Proper function of the Gastric Intestinal tract requires a delicate balance of healthy micro flora (probiotic bacteria) to support immunity, digestion, and nervous system responses. It is important that both beneficial and harmful microorganisms inhabit the human intestinal tract simultaneously. They enjoy a complex symbiotic relationship with each other. In fact, in many cases each contributes to the overall function and health of the intestinal tract, while keeping the other in check at the same time.

For example, even unfriendly microorganisms, such as the small colonies of candida yeast that inhabit the intestinal tract, carry out important tasks relating to human digestion. They aid in the digestion of sugars and release vital enzymes, nutrients and other essential substances as by-products of their work. Friendly micro-organisms keep candida yeast in check and prevent it from colonizing the digestive tract or allowing the candida yeast to grow uncontrollably.

While practices, such as antibiotics, are effective in killing germs and disease, they are also effective in killing beneficial bacteria. By introducing friendly microbes to the gut flora it can strengthen the resident bacteria. Although, probiotics have not proven to be effective at taking up residence themselves, they can improve the resident friendly microbes over a period of time.

We have all heard the saying "Death begins in the colon". Then it should come as no surprise that many Alternative Doctors believe that a **disrupted ecology of the gastrointestinal tract may be at the heart of up to ninety percent of all known human illness and disease**. The gastrointestinal tract's balance of beneficial flora is most commonly disrupted by antibiotic usage, excessive sugar consumption and stress, drinking chlorinated water, excessive alcohol consumption, frequent use of over the counter medication as well as prescription of anti-inflammatory drugs, painkillers and frequent consumption of colas or other carbonated beverages.

A diet high in red meats or rich, fatty foods will dramatically alter the acid/alkaline balance of the intestines, leading to overgrowth of disease causing, putrefactive bacteria that eventually overcome the beneficial bacteria and open the door to an onset of serious health problems. Moreover, colonies of putrefactive bacteria often discharge highly toxic by-products while reacting with foods in the digestive tract. This reaction could further upset the ecology of the gastrointestinal tract and slowly poison the entire body. The end result is the onset of chronic degenerative diseases.

This situation can be prevented through supplementation of the diet with foods rich in beneficial bacteria. Foods such as cultured yogurt, buttermilk, cottage cheese, whey and other soured milk food products are good examples of probiotic supplements.

According to Persian tradition, "Abraham" of the Old Testament owed his longevity to the ingestion of fermented milk and "King Francis I" of France was reportedly cured of an illness after eating yogurt in the early 1500's.

In 1908 Nobel Prize Winner, Ellie Metchnikoff studied the phenomenon of an incredible amount of people in Bulgaria living to be over 100 years old. He attributed their health and longevity to a microbe in the widely eaten Bulgarian yogurt, and he named the yogurt culturing microbe Lactobacillus Bulgaricus.

The scientific names of microbes always include the genus and species names and sometimes include a strain. Microbes considered to be beneficial to the human body include the genus names: Lactobacillus, Streptococcus, Bifidobacterium, and Sacchsromyces. Specific microbes include: Lactobacillus bulgaricus, L.acidophilus, L.casei, L.rueteri, Streptococcus latis, S.citrovorus, Bifobacterium bifidium, Saccharomyces boulardii and others.

Some of the benefits of probiotics may include: enhanced bowel function, prevention of colon cancer, improving of cholesterol, high blood-pressure, immune function, reducing infections, and inflammation, improving mineral absorption, preventing growth of harmful bacteria, fighting off diseases like candida, eczema and many more.

Food Fermentation

The mystical and chemical changes (biological transformations) that take place during the ancient practice of fermentation are well documented. Food fermentation is a process in which raw materials are converted to fermented foods by the growth and metabolic activities of desirable micro-organisms. Fermentation is the oldest known form of food biotechnology and causes enrichment of nutritive content by microbial synthesis of essential vitamins and improving the digestibility of protein and carbohydrates. There exists substantial tradition and science supporting the use of fermented and probiotic whole foods for the nourishment of mankind and it has been practiced throughout history.

Worldwide – alcohol, wine, vinegar, olives, yogurt, bread and cheese.
Asia – kombucha, miso, nata, amazake, sake, soy sauce and tofu.
Central Asia – kumis (mare milk), kefir, shubat (camel milk)
India – dosa, dahi (yogurt), idli (mixed pickle), u-to-nga
Africa – fermented millet porridge, oilseed, hibiscus seed and hot pepper sauce.
Americas – pickled vegetables, sauerkraut, chocolate, vanilla, tabasco.
Middle East – kushuk, lamoun, thorshi, boza.
Europe – sauerkraut, salami, prosciutto, cultured milk products, kefir.
Oceania – poi, kaanga pirau (rotten corn), sago

The Magic of Poi

For centuries, taro has been a nutritious staple food for Hawaiians and many other Polynesian peoples throughout the Pacific Rim. The Hawaiian's lived primarily on poi (fermented taro), sweet potatoes, fish, seaweed, coconut, green vegetables and fruit. Their diet contained no grains or animal milk. Very few food staples exist anywhere in the world that can be considered hypo-allergenic, are rich in calcium, potassium, phosphorus, magnesium, a good source of B vitamins, contain vitamins A and C, minerals, and a small measure of high quality, easily digestible protein.

Kalo, or Taro, as it is more commonly known, is not only nutritious but is considered to be a beneficial carbohydrate food that provides a high fiber, slow-release energy food source that may be beneficial for the more than 100 million people worldwide suffering from diabetes. The making of poi is indigenous to Hawaii and is the only place in the world where this ancient practice of food preparation is still found as a mainstream cultural and spiritual experience and may be successfully used by people with food allergies, food sensitivities, autism, celiac disease and various other health conditions that require a low-allergenic, easily digestible, food source.

Chocolate Grows on Trees

Chocolate – comes from the seeds of the cacao or chocolate tree. Theobromacacao Linnaeus ascribed the genus name meaning food of the gods to the Greek words, Theos (god) and broma (food). The chocolate tree produces large pods containing 30-40 bitter seeds embedded in a sweet sticky pulp. The sticky pulp was a staple food of the Mayans 2000 years ago and both the Mayans and Aztecs also used the seeds to make a chocolate drink. Once the pods are split open and the cacao beans removed, they are then fermented, dried, and roasted to develop the bean's flavor. Fermentation activates enzymes that create the beginnings of chocolate as we know it.

Food Cravings

If that craving for chocolate or sweets sometimes feels like it is coming from the depth of your gut that is because it may be. A small study links the type of bacteria living in people's digestive system to a desire for chocolate. Everyone has a vast community of microbes in their guts, but people who crave daily chocolate show signs of having different colonies than people who are immune to the chocolate's allure.

Crazy for sugar, or refined carbohydrates, comes from yeast overgrowth, otherwise known as candida. Blooms of candida yeast occur when your friendly intestinal bacteria are no longer able to keep candida under control like they are supposed to.

Sugar and other refined carbohydrates feed candida and as strange as it may seem, your health depends on the active population of friendly bacteria in your GI tract. At its optimum, this population includes about 500 kinds of bacteria, totaling at least 5 pounds of bacterial cells. The idea could eventually lead to treating some types of obesity by changing the composition of the trillions of bacteria occupying the intestines and stomach. This may be the case for other food-cravings too.

Research has shown that bacteria content in the intestines is different depending on body weight. **Normal weight people have one kind of dominant bacteria, while obese people's intestines are populated by a different strain**. It is still to be determined if the bacteria cause the cravings, or if early in life people's diets changed the bacteria, which then reinforced food choices.

The easiest way to encourage the growth of good bacteria is to add probiotics, fermented foods and drinks to your diet. Faced with the choice of assuming that intestinal bacteria are seizing control of their host's nervous system so as to relay their choice of meal preferences to the brain, and so control what the host eats, or the far simpler idea that the bacterial colonies thrive or perish based on what the host regularly eats. It seems to prove the point that bacteria changes in people who lose weight and they would therefore not make the same meal choices.

In the world outside the gut the flora might feed on what is available, with the numbers and makeup of population determined by the available food supply and the local environment's carrying capacity. Why do some foods like chocolate, wine and cheese taste so delicious? Fermenting magically transforms their original ingredients into something more desirable. Besides improving flavor, some lactic acid ferments, such as homemade sauerkraut, actually strengthens the immune system.

To the great complexity of the biological functions of the body belongs also its capacity of adaption. A healthy body can adapt itself to different types of nutrition and absorb the necessary minerals, vitamins and enzymes.

Chapter 9
THE DIGESTIVE TRACT

Digestion, Absorption and Elimination

During a person's lifetime, more than 30 tons of food pass through the digestive tract. When the gut (the small intestines) falters, disease and rapid aging is ensured and the origins of almost all chronic inflammatory disorders can be traced back to defects in the gut. Good health can only take root in good digestive functions.

The digestive tract is vitally important for total body health and must be healthy or the rest of the body suffers. It is 25-30 feet long, starts at the mouth and ends at the anus. The gut's function includes the digestion of the foods we eat into microscopic particles and the absorption and conversion of these particles into energy. The digestive tract has three phases: breaking down food, absorbing nutrients, and eliminating waste.

Digestion is a complex process turning food we eat into the energy we need to survive. When food is consumed, digestion starts in the mouth, where starches are broken down into simple sugars by the action of enzymes. It cannot be emphasized enough to chew food well. Benefits of proper chewing include:

Better immunity – alkalizing food, enzymes break down nutrients and toxins, strengthening of jaw muscles (stomach meridian begins in the face), helps complexion by increasing circulation.

Calm nervous system – slow chewing is centering and grounding, very meditative which brings a more liquid/aware state. Food tastes better (especially starches).

Head/stomach connection – better chewed food equals less work for the stomach. As food is swallowed, it arrives at the stomach, which secretes acids and enzymes to help break down proteins and mechanically churn and break the food further down into liquid form.

The Chinese say the stomach ripens ingested food, sends the purest portion to the spleen to be made into blood and life-force, and sends the more turbid or indigestible parts to the small intestines for further digestion. Raw food sits in the upper stomach until it is digested in its own juices, every raw food possesses enzymes to digest itself. Digestive enzymes help us break down and absorb food, while systemic enzymes help with bodily functions like building new tissue, making blood and immune cells. Enzymes are involved in almost every cellular activity.

Essence extraction function – unknown to western medicine. Sun-essence – chlorophyll is the plant's blood, produced with sunlight, water, minerals, chloroplast cells, and microbes. The Chinese medicine texts say the stomach absorbs the finer essences in food, the more subtle essences are certain salts, water (which is an incredible carrier of information), alcohol (from liquor or other food/beverages), and sugars. One can feel satisfied from the stomach usually before it is actually full, if one is eating life and nourishing foods and chewing slowly.

Importance of stomach acids and mucous – it is very important that our stomach produces enough acid, this acid makes protein and vitamin B12 available, kills unfriendly bacteria and causes mucous to be produced to protect the stomach lining.

Liver/spleen/gallbladder – the liver deals with all digested foods. All blood vessels in the gut go to the liver before any other body system because the liver has the ability to break down harmful substances that can be produced from digested food, and from hormonal and microbial by-products. We cannot process fats easily without bile, which is made from cholesterol. Bile is an acid and acids are irritating to tissues. This irritation is helping to move food along the intestines for excretion.

Small intestines – separate the impure from the pure – this area is where most of the physical absorption and assimilation occurs. **Food may be the single biggest challenge to our immune system**. Because our intestines are lined with lymph and immune tissue they protect the delicate environment from potential invaders that can be absorbed through the gut wall (leaky gut) and bypass the liver system. When blood, fluid and mucous get stagnant in the digestive tract, blood flow slows to the head and brain and may cause foggy thinking and delusional decision making.

An intelligence also resides in the digestive tract, it is our microbial balance and enteric brain. If our intestinal flora is balanced, the lining of the gut wall is strong and our **body's ability to discern what is self from non-self is good.**

One of the most important factors for absorption and utilization of the energy and elements of foods is preparing the food so it can be assimilated. Pre-digestion can be soaking our grains, beans, nuts and seeds for 24 hours, culturing and cooking cabbage family and oxalic vegetables (like beets, chard and spinach), culturing pasteurized milk (kefir, yogurt and cheeses), and marinating meat and fish.

It is unknown whether the knowledge of soaking grains, culturing dairy and vegetables, processing and pre-treating meat, and other food processing methods were specifically designed for digestion, or if these techniques could have had beneficial side effects. The main goal of ancient people was probably food preservation.

Large intestines/colon – The turbid or least nutritionally active portion of the food finishes its journey in the large intestines, where the largest amount of microbes in our digestive tract go to work on it. Our body will absorb water, salts, and some vitamins, which are produced by our helpful gut bacteria. Our colon is very important in keeping us hydrated and in helping us manage the mixture of partially digested food.

Here the body is extracting water from the mostly digested food as it is pushed and squeezed along. Microbes feed off this waste. Many famous healers and healing traditions, including Ayurveda have mentioned the importance of a clean colon. This is because toxic water can be reabsorbed, potentially clogging the blood vessels leading back to the liver or the lining of the colon can become congested with mucous as a defense mechanism against the overly toxic substances.

Here our intestinal microbes can make vitamin B's, transform hormones, kill and crowd out unfriendly bacteria and break down toxic material. There is more microbial activity in the colon than any other place in the gut. The good bacteria in this area is essential to prevent the transformation of carcinogenic chemicals into cancer causing agents.

The final stop for the now firm digested material is the rectum, just above the anus. The stool sits in this area until a signal reaches the brain to let go. In macrobiotic studies, there is a correlation between the mouth and the anus. One is food's beginning journey into the body (becoming part of us) and the other end is where the discarded parts of the body and food leave to help fertilize the soil, the earth. The ending is also the beginning of a new cycle of life. No matter what, we are tied to the cycles of nature. **When we eat, we are not only feeding our own cells, but the bacteria we share our bodies with.**

The health of the human digestive tract is a much more important and determinant of vitality than most people realize. Though it runs deep through the center of the body, the digestive tract is actually exposed to the outer environment through the intake of food.

To help us digest and process food, promote regularity and help with mineral absorption, added probiotics may support the small intestine to promote healthy immunity, inhibit yeasts, enhance digestion of milk products and may also benefit the large intestine as the first-choice to regulate proper elimination.

Using probiotics for constipation is the premise that supplying the gut with healthy bacteria in the form of probiotics can help to normalize intestinal function. By promoting normal digestive function, it is thought that bowel motility will be enhanced and the symptoms of constipation relieved.

Chronic Constipation

Chronic constipation is no laughing matter. Not only can it be an uncomfortable condition causing such symptoms as abdominal bloating and bad breath, long term constipation may increase your risk of contracting colon cancer. Of late, there has been an increased interest in using probiotics to treat a variety of intestinal disorders including diarrhea and constipation.

Constipation is defined as having a bowel movement fewer than three times per week. With constipation stools are usually hard, dry, small in size, and difficult to eliminate. Some people who are constipated find it painful to have a bowel movement and often experience straining, bloating, the sensation of a full bowel and possibly hemorrhoids.

Intestines are lined with microbes that help digest food. This microbial lining or intestinal flora, which adds a large amount of microbes to the fiber that passes through the intestines providing bulk and help to retain moisture. Without these microbes the urge to pass the bowel contents in a timely manner will be difficult and will consequently cause constipation. Originally, the milk-microbes acquired through breast-milk are supporting the growth of the intestinal flora, but need to be nurtured and maintained.

However, for the last 50 years we have been sterilizing our soil with pesticides and herbicides, destroying most bacteria, both bad and good. Our modern lifestyle, which includes antibiotic drug use, chlorinated water, chemical ingestion, pollution, and poor diet, is responsible for eradicating much of the beneficial bacteria in our bodies and it is the lack of beneficial microbes that often results in constipation, excess gas and bloating. In addition, these microbes work from the inside of the intestines, dislodging accumulated decay on the walls and flushing out this old waste.

Microbes work in symbiosis with somatic (tissue or organ) cells to eliminate toxic waste that has built up from years of constipation. Besides laxatives, stool softeners, enemas and colon therapy, the ancient practice of assuming a natural position is a long forgotten practice in the western world.

The Natural Position

Using the squatting position to evacuate waste is a natural constipation remedy that many doctors do not discuss with their patients. Thus many who are now suffering from constipation will not be able to obtain permanent relief and go through life not knowing that the underlying cause of their constipation may be their toilet posture.

Squatting can prevent and cure constipation in many ways:

1) In the squatting position, the entire weight of the upper body rests on both feet and presses against the thighs. This naturally supports and compresses the colon which, along with gravity, enables waste to be expelled more easily.

2) Squatting protects the nerves that control the prostate, bladder and uterus from becoming stretched and damaged.

3) Securely seals the ileocecal valve, between the colon and the small intestines.

4) Relaxes the puborectalis muscle, which normally chokes the rectum in order to maintain continence.

The positions and modalities of defecation are culture dependent. The natural and instinctive method used by all primates, including humans for defecation, is the squatting position. Squat toilets, sometimes referred to as 'natural-position toilets' are still used by the vast majority of the world, including most of Africa and Asia.

The widespread use of seated-position toilets in the Western World is a recent development, beginning in the 19th century with the advent of indoor plumbing.

Bockus in Gastroenterology, (the standard textbook on the subject) states: The ideal posture for defecation is the squatting position, with the thighs flexed upon the abdomen. In this way the capacity of the abdominal cavity is greatly diminished and intra-abdominal pressure increased, thus encouraging expulsion.

Diarrhea

Among healthy individuals the maximum number of daily bowel movements is approximately three, diarrhea can be defined as having more bowel movements than usual. Thus, if an individual who usually has one bowel movement each day begins to have two bowel movements, then diarrhea is present.

If there is inflammation or damage in the gut, water cannot be absorbed and diarrhea may be experienced. Viral infections and bacterial toxins are the most common causes of diarrhea, though stress and diet can contribute to an inflamed colon as well.

One pathogenic bacteria in particular, Clostridium difficile, thrives in the gut after taking anti-biotics. Clostridium difficile causes inflammation that leads to diarrhea in many people and is often found in hospitals.

Fecal Transplants

In 2008, Dr. Alexander Khoruts, a gastroenterologist at the University of Minnesota ran out of options to treat one of his patients suffering from a vicious gut infection of Clostridium difficile. Crippled by constant diarrhea, which left her in a wheelchair wearing diapers, his patient was just wasting away, losing 60 pounds over the course of eight months. Treating her with an assortment of antibiotics, could not stop the bacteria. Dr. Khoruts decided that his patient needed a transplant, not a part of someone else's intestines, or a stomach, or any other organ. Instead, he gave her some of her husband's bacteria.

After mixing a small sample of her husband's stool with saline solution and delivered it into her colon, Kharuts and his colleges reported that her diarrhea vanished within a day. Her C. difficile infection disappeared as well and has not returned since.

The procedure – known as bacterio-therapy or fecal transplantation had been carried out a few times over the past few decades, but Khoruts and his colleges have been able to do something that other doctors could not do during a fecal transplant: They took a genetic survey of the bacteria of their patients intestines before and after the transplant to later be able to track the beneficial microbes.

Before the transplant, they found, her gut flora was in a desperate state. The normal bacteria just did not exist – she was colonized by all sorts of misfits. Two weeks after the transplant, the scientists analyzed the microbes again. Her husband's microbes had taken over and that community was able to function and cure her disease in a matter of days. Since then additional transplants have been performed, many of which cured their patients.

The unusual treatment has a definite ehhww factor. First the donor feces is stripped of larger particles and then blended with a saline solution. The resulting mixture is transferred to the recipient via an enema – or even through a tube inserted into the mouth or nose.

C.difficile, is increasingly common and often hard to treat with antibiotics. Once infected, it is notoriously difficult to get rid of it, resulting in copious and sometimes frequent diarrhea, and occasionally a more serious and painful condition called colitis (inflammation of the colon). Benefits of fecal bacterio-therapy include reducing the risk of cultivating antibiotic associated resistance in the pathogenic bacteria displaced by the colonization and its purported effectiveness when antibiotic resistance is already in place.

Scientists are continuously amazed by the diversity, power and sheer number of microbes that live in our bodies. Several teams are now working together to unravel this complexity in a systematic way, known as the Human Micro-biome Project.

For more than a century, treating patients with beneficial bacteria and probiotics, have had only limited success. The problem may lie in our ignorance of precisely knowing how these microbes in our bodies affect our health and a better understanding of the micro-biome might give the medical profession a new way to fight some of the diseases.

Chapter 10
GUT – BRAIN CONNECTION

Digestion and Mental Health

It is through our digestive system we obtain the nutrients our body needs to survive. Vitamins, minerals, amino acids, fatty acids and many more are primarily provided from the foods we eat and the fluids we drink. However, when our diet is lacking and / or our digestive system is not working optimally, nutrient deficiencies and a range of inflammatory immune responses will occur. This can lead to a whole array of health problems including mental disturbances such as depression, anxiety and even schizophrenia.

Studies show that people with digestive-based problems and diseases (e.g. celiac disease, Crohn's disease, irritable bowel syndrome) are significantly more likely to suffer from mental health problems such as major depression, anxiety, and somatoform disorders (mental disorder characterized by physical symptoms that mimic physical disease or injury for which there is no identifiable physical cause) occur in up to 94% of people with irritable bowel syndrome.

Unfortunately the gut brain connections and digestive disturbances are regularly overlooked in mental health treatment. ADD, ADHD, Anxiety,

Autism, Depression, OCD, Bipolar Disorder, even rheumatoid arthritis and Celiac disease are thought to be related in that they all originate in the gut from poor gut flora.

Some of the common causes of digestive problems that require assessment and treatment if present with mental health problems and digestive complaints are:

Food allergies / intolerance – for many people a specific food allergy and depression are strongly interlinked.

Medication – many medications cause digestive problems. For example, prolonged use of non-steroidal anti-inflammatory drugs (NSAIDs, medications containing ibuprofen) and aspirin, oral contraceptives, anti-depressants, cholesterol lowering drugs, chemo-therapeutic drugs, diuretics and blood thinning drugs. In particular medications that reduce acid levels in the gut.

Parasites and pathogens – can cause havoc on the digestive and immune system.

Digestive enzyme deficiencies – linked to properly break down fats, proteins and starches and the ability to absorb nutrients.

Stomach acid problems – Stomach acid (hydrochloric acid or HCI) is excreted by specialized cells to help our body digest protein and minerals, and to sterilize foods.

Bacterial imbalances (dysbiosis) – Our digestive system contains billions of bacteria (gut flora) that have many crucial roles in our body including pathogen defence, digestion, and synthesis of vitamins.

Intestinal permeability (leaky gut) – Our

gastrointestinal system has a mucosal barrier that protects our internal system (e.g. heart, lungs, brain) from outside world pathogens. This mucosal barrier consists of tightly joined cells (tight junctions) that show certain nutrients to be absorbed. When disruption in these tight junctions occurs, many dietary and bacterial substances and other toxic by-products can leak into the bloodstream leading to an immune reaction causing an array of symptoms some of them related to mental functions.

Excessive, prolonged stress – can have a negative effect on our digestive system and can impair healing. Excess stress can also lead to poor lifestyle and dietary habits that can exacerbate digestive problems.

Dr. Michael Gershon states in his book "The Second Brain" that **we have more nerve cells in our bowel than in our spine and our brain and since the enteric nervous system can function on its own, it must be considered possible that the brain in the bowel may also have its own psychoneuroses.** In other words, our gut can operate independently from the brain – running processes we may not be aware of.

Insanity and Mental Illness

Microbes are the greatest predator to man. As medical technology improves, there is increasing recognition that infectious disease contributes not only to acute, but also to chronic conditions, relapsing illness and mental health. The evidence to support this is a combination of insights from theoretical biology (particularly Darwinian medicine), research, and direct clinical observations.

The most common sequence of disease begins with a vulnerability and an exposure to one or more stressors and may include genetic and/or increased sensitivity as a result of chronic stress. Although infection may occur from microbes that are always present in the environment, a greater number of organisms or more virulent forms can further increase risk. The course of infection most relevant to psychiatry includes injury from a prior infection, chronic, low-grade, persistent relapsing infections, or the persistence of the infectious agent in the inactive state.

Psychiatric syndromes caused by infectious disease most commonly include depression, obsessive compulsive disorder (OCD), panic disorder, social phobias, variants of attention deficit disorder, episodic impulsive hostility, bipolar disorders, eating disorders, dementia, various cognitive impairments, psychosis, and some cases of dissociative behaviour.

In clinical experience, the link between infectious disease and psychopathology has been an issue with Lyme disease, syphilis, babesiosis, ehrlichiosis, mycoplasma pneumonia, toxopiasmosis, stealth virus, borna virus, AIDS, CMV, herpes, strep and other unknown infectious agents.

It is a theory sharply at odds with earlier views of the genesis of psychological illness. Followers of Freud long held that mental and emotional trouble is primarily the result of poor parenting, especially by mothers. But such theories, which added immeasurable guilt to parents with mentally ill offspring, have turned out to have little evidence to back them up, most experts now agree.

Instead in recent years, the focus has shifted to genes as the main source of mental illness. Faulty DNA is thought to be at least partly responsible, yet genetics does not appear to wholly account for the occurrence of major psychiatric ailments.

If heredity alone were to blame, identical twins would develop schizophrenia with a high degree of concordance, but in fact only 40% of cases in which one identical twin has the disease does the other twin have it as well. Autism, though it has been observed to run in families, also strikes five out of every 10.000 children apparently arbitrarily, nor can depression and other affective disorders be completely explained by damaged DNA. Bacteria may be the key.

Since we know so little about the viruses and bacteria that cause some types of mental illness, it makes sense to avoid them when we can. Schizophrenia is a devastating illness. With one percent of the world's population suffering from its symptoms of hallucinations, psychosis and impaired cognitive ability. The disease destroys relationships and renders many sufferers unable to hold down a job.

What could cause such frightening damage to the brain? According to a growing body of research, the culprit is surprising – **the flu**.

In summary, the complexities of these issues teach us humility. To better understand the clinical syndrome associated with these infections, we need to recognize the significance of mental symptoms in chronic interactive infections and psychiatrists need to better appreciate the role of microbes in causing mental illness.

Chapter 11

MICROBES AND DISEASES

Connection to Physical Health

Looking at live blood through a microscope, we can see a reflection of cause and effect. Whatever the current metabolic condition, the internal microbes will co-exist with us and will be in perfect balance with whatever environment is provided. The blood will reflect our state of physical, emotional and mental health.

Interestingly, by energetically removing some stressors and changing the internal environment, blood has the ability to respond almost immediately by adapting to a new state of health.

Blood bacteria are thought to be connected with the origin of life. Livingston (1906-1990) believed these microbes were responsible not only for the initiation of life, but also acted as terminators leading to death, admittedly a difficult concept for most people to consider. Wilhelm Reich (1897-1957) referred to bacteria emanating from energy-depleted cells as T-bacilli (T-derived from the German word "Tod" – meaning death). He found T-bacilli in both healthy and sick individuals. However, in the blood of sick people they were more numerous.

Several assumptions are utilized to facilitate the presentation that the expressed microbiological form in the blood of a healthy individual are predominantly that of one micro-organism with a capability of different formats of expression.

Despite a century of modern medicine we know little about the cause of cancer and the many chronic diseases that accompany old age. Heart and blood vessel disease (arteriosclerosis) are the most common causes of death in the elderly. Could blood bacteria contribute to the cellular changes in the heart and blood vessels?

Pleomorphic bacteria has a life cycle and so do we. We ourselves are pleomorphic in that we began life as microscopic beings and grow to produce new life by mixing our genetic material with others. When we die, we hope to continue as spirit with eternal life. In his experiments, Wilhelm Reich was astonished to discover that it was impossible to destroy the smallest living forms of life – the microbe.

All human blood is infected with bacteria and the microbiology of the blood is intimately related to the proposed cause of cancer. Continuing research dating back to the late nineteenth century indicates that pleomorphic (variably appearing) bacteria are implicated in cancer and over the past few decades more and more studies have confirmed that similar bacteria can be found in the blood.

Controversial researchers who made outstanding contributions to the study of pleomorphic microbes in human diseases include Antoine Bechamp, Gunther Enderlein, Wilhelm Reich, and others.

Cancer Connection

Researchers have known for decades that the bacteria harbored in our bodies are not innocent bystanders but rather active participants in health and disease. Yet only recently have molecular methods evolved to the point that they can identify and characterize some of our microbial residents.

The methods used to determine the different bacterial groups contained within biopsies from 45 patients undergoing colonoscopies uncovered a higher bacterial diversity and richness in individuals found to have adenomas than in those without these colorectal cancer precursors. In particular, a group called Proteo-bacteria was in higher abundance in these cases than in controls, which was interesting considering that it is the category where E.coli and other common pathogens reside.

It is still not clear whether alterations in bacterial composition cause adenomas, or if adenomas cause the altered balance. The ultimate goal may be to determine if the differences found in the mucosa lining in the colon also exist in the luminal or fecal matter that passes through the colon.

By looking at bacteria and their role, it opens up a whole new world and gives us a better understanding of the entire spectrum of factors involved in cancer – diet, environment, genes and microbes.

Multiple Sclerosis

Biologists at the California Institute of Technology have demonstrated a connection between multiple sclerosis (MS) – an autoimmune disorder that effects the brain and spinal cord – and gut bacteria. Although the cause of MS is unknown, micro-organisms seem to play some sort of role. For example, the disease gets worse after viral infections, and bacterial infections cause an increase in MS symptoms.

This study shows for the first time that specific intestinal bacteria have a significant role in affecting the nervous system during Multiple Sclerosis – and they do so from the gut, an anatomical location very far from the brain.

Perhaps treatment for diseases such as multiple sclerosis may someday include probiotics that can restore normal immune function in the gut – and the brain.

Migraines

There is an ever increasing level of information pointing to serotonin and its production and/or regulation as playing a significant role in migraines. Furthermore all the neurotransmitters that regulate brain activity also play important roles in the enteric nervous system, creating an obvious connection to the gastrointestinal aspect.

Whether we look at the nausea they may suffer, the acknowledgment that certain foods, drinks or additives can trigger migraines or the recent research on gastric disturbance of homeostasis in migraines one cannot avoid the fact that there is a gut brain connection. Not only is it possible that signal substances like serotonin and dopamine are subject to regulation by bacteria, synapse function may also be regulated by colonizing bacteria.

Chapter 12
ANTIBIOTICS

Mold and Penicillin

We have not always relied on the latest new medicines to remedy what ails us. At one time, we used what we had around us, even if it was something as primitive or supposedly disgusting as a piece of moldy bread, urine or honey, among others. Egyptians used honey for wound dressing and ancient Indian healers reportedly used urine as treatments with great success. As mould is the father of modern antibiotics, we shall focus on the example of moldy bread and anti-bacterial benefits in general.

Perhaps one of the earliest recorded examples of natural antibiotics was in ancient Serbia, where old bread was pressed upon wounds to help prevent infection which contained an early, raw form of penicillin. The mold killed off festering infections, and the technique became more widely known. Mold-use was also practiced in ancient China and Greece, much the same way as the Serbians used it. Even the Sri Lankan army in the first century BC used mold-based treatments for their wounded.

However, with the mass-production of bio-chemical antibiotics that followed successful research allowed for penicillin to be produced in quantities large enough that animals could begin to be used in experimentation, and eventually lead to the large scale manufacturing in 1941.

Over time some strains of bacteria began to adapt and became resistant to antibiotics. Today, not all antibiotics work for all bacteria, due to acquired resistance. Inappropriate antibacterial treatment and overuse of antibiotics have contributed to the emergence of antibacterial-resistant bacteria. Common forms of antibacterial misuse include excessive use of antibiotics, failure to take the entire prescribed course, incorrect dosage and administration, or failure to rest for sufficient recovery.

Another inappropriate use is the prescription of anti-bacterials to treat viral infections, such as the common cold. Use of antibiotics as growth promoter in agriculture given to livestock in the absence of disease are additional examples of misuse – farmers need to restrict adding it to animal feed.

The successful outcome of antimicrobial therapy with antibacterial compounds depends on several factors and include host defense mechanisms, the location of infection, and the pharmacokinetic and pharmacodynamic properties of the antibacterial. Adverse effects may range from fever and nausea to major allergic reactions including photo dermatitis and anaphylaxis.

Common side effects include diarrhea, resulting from disruption of the species composition in the intestinal flora, allowing, for example, an overgrowth of pathogenic bacteria, such as Clostridium difficile.

Chapter 13
EXPOSURE TO BACTERIA

Are Sterile Environments Making Us Sick?

Is the pursuit of a bacteria-free world making us sick? The rising incidence of asthma and allergies in the developed (cleaner) world could be tied to the relatively sterile environments our children live in compared to a generation ago. Children not exposed to harmful bacteria, or conversely, given antibiotics to kill bacteria, do not receive the germ exposure required to make antibodies.

More specifically, they do not develop T-helper cells, which fight foreign cellular invaders and minimize allergies. Children exposed to endotoxin-releasing bacteria, for example, are less likely to be allergic to cats and dogs. Ridding ourselves of bacteria is a hopeless endeavor.

For several years scientists have uncovered signs that illness can result when the immune system does not get enough practice fighting bacteria and viruses. Several epidemiological and experimental studies have converged to put this hypothesis on firm ground. Although it had been thought that the immune system required periodic infections during childhood, researchers now argue that the main problem is that the children have become squeaky clean. They suspect that children need contact not with disease-causing microbes, but with bacteria in soil and untreated water, particularly with organisms called mycobacteria, to strengthen their immune system.

The continuing rise in allergies coincides with the acceptance and use of antibacterial soaps and hand sanitizers, the ultimate excessive cleaning. Antibacterial soaps are applied to the hands and bodies in the shower or during hand washing with warm water, allowing them to easily enter the body while pores are open. There is gathering evidence that these anti-bacterial ingredients are also carcinogenic. Washing hands the old fashioned way, with regular soap and water will send the bacteria down the drain and no harmful substances enter the body's bloodstream.

Chapter 14
ANCIENT REMEDIES

Poisons, Spells and Medicine

Egypt – If one had to be ill in ancient times, the best place to do so would probably have been Egypt. Not that an Egyptian's chances of survival would have been significantly better than those of his foreign contemporaries, but at least he had the satisfaction of being treated by physicians whose art was renowned all over the ancient world. Unlike the injuries by accidents of fighting, which were dealt with by the *zwn.w* or scorpion stings and snake bites for which the *xrp srqt,* the exorcist of *Serqet,* knew the appropriate spells and remedies, illnesses and their causes were mysterious.

The Egyptians explained them as the work of the gods, caused by the presence of evil spirits or their poisons, and cleansing of the body was the way to rid the body of their influence. Incantations, prayers to the gods – above all to Sekhmet, the goddess of healing, curses and threats, often accompanied by the injection of nasty smelling and tasting medicines into the various bodily orifices, were hoped to prove effective.

Preventive measures included prayers and various

kinds of magic and above all the diversity and rich well-irrigated soil resulted in a diet which was reasonably balanced and provided carbohydrates from cereals, vitamins from fruit and vegetables, and protein mostly from fish. Milk and milk products were just occasionally consumed, as were legumes, seeds and oils. The idea that all diseases and death begin in the colon is one of the oldest health concerns known to humankind. The ancient Egyptians associated feces with decay, and decay with health. This caused them to write in ancient papyri that decay began in the anus. The Egyptians were obsessed with preserving corpses. Embalmers observed the petrification by bacteria (a normal process within the intestines after death) and followed the practice of removing the stomach and intestines as part of the embalming process. Worry about decay governed their life and constipation was to be avoided.

Roman Medicine – Ancient Roman medicine was a combination of some limited scientific knowledge, and a deeply rooted religious and mythological system. While knowledge of anatomy was quite impressive, and many surgical techniques were only surpassed in the modern age, the application of medicines and cures was simplistic and largely ineffective. Much of the Roman system was adopted from the Greeks, and primarily the teachings of Hippocrates.

Hippocrates (460-384 BC), is largely recognized as the father of modern medicine, as he created the concept of medicine in a separate scientific field away from philosophical and mythic approach. The original Hippocratic Oath . . .

I SWEAR by Apollo the physician, Aesculapius, and Health, and All-heal, and all the gods and goddesses, that, according to my ability and judgement, I will keep this Oath and this stipulation. TO RECHON him who taught me this Art equally dear to me as my parents, to share my substance with him, and relieve his necessities if required; to look up his offspring in the same footing as my own brothers, and to teach them this art, if they shall wish to learn it, without fee or stipulation; and that by precept, lecture, and every other mode of instruction, I will impart a knowledge of the Art to my own sons, and those of my teachers, and to disciples bound by a stipulation and oath according the law of medicine, but to none others.

I WILL FOLLOW that system of regimen which, according to my ability and judgment, I consider for the benefit of my patients, and abstain from whatever is deleterious and mischievous. I will give no deadly medicine to any one if asked, nor suggest any such counsel; and in like manner I will not give a woman a pessary to produce abortion. WITH PURITY AND WITH HOLINESS I will pass my life and practice my Art. I will not cut persons laboring under the stone, but will leave this to be done by men who are practitioners of this work. Into whatever houses I enter, I will go into them for the benefit of the sick, and will abstain from every voluntary act of mischief and corruption; and, further from the seduction of females or males, of freemen and slaves.

> *WHATEVER, IN CONNECTION with my professional practice or not, in connection with it, I see or hear, in the life of men, which ought not to be spoken of abroad, I will not divulge, as reckoning that all such should be kept secret.*
>
> *WHILE I CONTINUE to keep this Oath unviolated, may it be granted to me to enjoy life and the practice of the art, respected by all men, in all times! But should I trespass and violate this Oath, may the reverse be my lot.*

. . . stems directly from Hippocrates and a modern version continues to be the binding ethical law guiding all those in the field of medicine. He was primarily responsible for the foundation of recording illnesses, attempts at treatment, and the causes and effects.

Despite the reliance on a mystical approach to healing, Roman society maintained reasonably good health throughout history. The exhaustive use of aqua-ducts and fresh running water, including toilets and sewer systems, prevented the proliferation of many standing water based diseases, and also washed away wastes from heavily populated areas.

Excellent hygiene and food supply also played a prominent role. The Roman baths were an integral part of society, in all social classes, and regular cleansing helped fight germs and bacteria. The Romans also tried, whenever practical, to boil medical tools and prevent using them on more than one patient without cleansing.

India – Ayurveda is considered by many scholars to be the oldest healing science, embracing a holistic approach to health that is designed to help people live long, healthy, and well-balanced lives. The term Ayurveda is taken from the Sanskrit words ayus, meaning life or lifespan, and Veda, meaning knowledge. It has been practiced in India for at least 5,000 years and has recently become popular in Western cultures.

The basic principle of Ayurveda is to prevent and treat illness by maintaining balance in the body, mind, and consciousness through proper drinking, diet, and lifestyle, as well as herbal remedies. It is believed that building a healthy metabolic system, attaining good digestion and proper excretion leads to vitality. Ayurveda also focuses on exercise, yoga, meditation, and massage. Thus, body, mind, and spirit/consciousness need to be addressed both individually and in unison for health to ensue.

China – Acupuncture treatments can be traced as far back as the Stone Age, when stone knives and other sharp-edged tools were used to relieve pains and diseases and were known by the ancients as bian (meaning stone to treat diseases). Later the bian stones were replaced by needles made of bone or bamboo and the theory of channels was developed which has been the guiding principle of acupuncture therapy for more than 5000 years. Recent research using scientific knowledge and methods, has shown that stimulation of the acupuncture points may produce various psychological functions in the body.

The Chinese believe that man is a small universe,

a microcosm of the large one we live in. Every physical and mental function is governed by the same universal laws of energy and opposing forces (yin and yang) that regulate the universe. Accordingly, everything is linked into a vast coordinating entity. The great leader of the Chinese people Chairman Mao Tsetung said: In the field of the struggle for production and scientific experiment, mankind makes constant progress and nature undergoes constant change.

Europe – Medieval medicine in Europe, up to the 19th century, evolved around the theory of humors and herbs. Humorism is based on the belief that health is driven by four different fluids (black bile, yellow bile, phlegm and blood) which influence our well-being. They have to be in perfect balance and are linked to seasons, organs, temper and elements. At that time, the link between bacteria and contagion was still unknown. Ignaz Semmelweis, a physician and his students would perform autopsies on bodies of dead women who were struck by dreaded disease. Later, with unclean hands, they would make pelvic examinations in the obstetrical ward and as a result, many of those examined, died. The hospital doctors attributed their deaths to constipation, delayed lactation, poisoned air and fear.

Semmelweis reasoned that some deadly substance was being transferred from his staff to the patients. He ordered all personnel to thoroughly wash their hands and the death rate was drastically reduced. He was faced with bitter opposition and contempt from his colleges at that time.

Americas – Native American healing has been practiced in North America for up to 40,000 years. It has been influenced by the environments in which Native Americans settled, and the nature, plants, and animals around them. Other healing practices were influenced over time by the migration and contact with other tribes along trade routes. The tribes gathered many herbs from the surrounding environment and sometimes traded over long distances.

The natives from the Arctic to Antarctic knew that humans are parasitized and harbor these invaders like any other animal. Purifying and cleansing the body is an important technique used in Native American healing. The ancient parasitic remedies used were frequent purging's that included diarrhea and vomiting to rid themselves of these invaders and many cultures continued these practices up to recent times.

Healers may include shamans, herbalists, spiritual healers, and medicine men or women. Many Native Americans see their healers for spiritual reasons, such as to seek guidance, truth, balance, reassurance, and spiritual well-being – they believe that the spirit is an inseparable element of healing. Because some illnesses are believed to come from angry spirits, healers may also invoke the healing powers of spirits and may use special rituals and herbal mixtures to try to appease these spirits.

Hebrew Scriptures – As reluctant as medical science has been in advancing effective methods for controlling contagion, it is utterly remarkable that 3500 years ago Hebrews practiced sanitary measures that were not only superior to those proposed by Semmelweis, but were even up to par with 21st century standards and are now being recognized by science.

Food – Hebrew Scriptures recorded dietetic principles with information to include the superiority of fruits, nuts, grains and green vegetables for optimum health. Later as flesh foods were adopted, there was outlined a detailed classification of meats fit and unfit for human consumption. The Hebrew's were allowed to eat fish with fins and scales, and herbivorous, ruminant animals. Birds of prey and carnivorous beasts were unclean in their eating habits, and therefore more susceptible to disease than clean domesticated stock.

Warnings were given against the consumption of animal fats and blood. Medical authorities today recognize that fat stored in a carcass contains a high concentration of toxic waste or parasites, and if there is any disease in a dead animal, it infects the blood throughout the body. (These principles were also outlined in Muslims and Christian scriptures)

Hygiene – While many nations around them suffered sicknesses aggravated by poor hygiene, the Ancient Hebrew Civilization practiced superior standards of cleanliness and the ancient rules governing antisepsis and sterilization were as modern as today.

Penicillin – About 1000 BC, David, King of Israel wrote: Purge me with hyssop, and I shall be clean: wash me, and I shall be whiter than snow. (Psalm 51:7) – Centuries before Pasteur, people used certain herbs for their anti-viral, anti-bacterial, and anti-biotic purposes; hyssop was one herb that contains penicillin as an active ingredient.

Chapter 15

NATURAL HEALING REMEDIES

The Silver Spoon

Most people know the saying "Born with a silver spoon in your mouth". This means that the child would never want for anything. Since the 17th century, the silver spoon symbolizes great fortune and privilege. Prosperous individuals would be presenting them to newborns at the christening.

The ancient civilizations of Greece and Rome have known for thousands of years that silver has anti-bacterial properties and used silver to control bodily infection and prevent food spoilage. Since before the birth of Christ, Greeks used silver vessels to prevent spoilage of liquids and the Romans, used silver urns for wines and similar purposes. Early American settlers dropped silver coins into their milk to kill bacteria and to keep it fresh until use.

Today it is still used widely in water filters and tanks. Silver kills bacteria, viruses, fungi, yeast and molds. It does this by degrading the enzymes these microorganisms need to produce cellular energy, it interrupts a bacteria's cell ability to form chemical bonds essential to its survival and ultimately suffocates the bacteria. Human cells have thick walls and are not disturbed by its presents. Silver is stable and has prolonged effectiveness and may be the answer to the overuse of antibiotics which is causing strains of microbes resistant to drugs.

Throughout the Middle Ages in Europe, it had been noticed that royalty, who consumed their food and drink from silver tableware, utensils and cups, tended to develop a bluish skin tone (Blue bloods). This was thought to be due to the silver that entered their bodies during consumption. Furthermore the lower rate of plague-related deaths among royalty led to the inference that silver could be protective.

During the 14th century in Europe about 25% of the total population died from the Bubonic Plague, which swept the continent. Only the Gypsies seemed to be immune from its ravages due to their practice of grinding up silver into small particles, then inserting tiny amounts into an open vein. The particles were distributed through blood circulation and successfully destroyed harmful bacteria and viruses.

Silver is a powerful, natural antibiotic and was used to prevent and treat infections in the late 19th and early 20th century for a variety of ailments. However, in the late 1930's, the medical profession turned its attention to pharmaceutical drugs and until the mid-70's use of silver in antimicrobial applications was largely ignored. Studies conducted over decades show that even at less than one part per billion in pure water, silver disrupts bacteria cells permanently.

Antibiotics on the other hand, like penicillin, are bacteriostatic – they temporarily stop the growth of bacteria. This action is faster than that of silver, but when the drugs cease their attachment to the bacteria, the bacteria reactivates and becomes resistant to the drug.

Contemporary medicine is in a quandary as our strongest antibiotics are ineffective against many strains of harmful bacteria. Physicians became aware again of the benefits of silver, and advised their patients (if they could afford it) to eat on silver plates and use silver spoons to stay healthy. People began to give their babies silver spoons to suck on, and forestall any illnesses. That is where the expression "Born with a silver spoon in your mouth" originated. Silver leaps into action when it comes across bacteria and gets to work to destroy it, thereby preventing infection and stopping bacteria from breeding.

Virtually anyone reading this probably had a drop of silver nitrate solution dropped into their eyes after birth, which became standard practice at the end of the 19th century to prevent blindness, in the event that the mother had a venereal disease.

Astonishingly about 150.000 Americans die every year from taking medications as prescribed. Many Americans are looking for answers on their own and are finding them in homeopathic remedies. Some medical doctors are taking a step back in time and getting excellent results in fighting hundreds of illness causing bacteria, viruses, and fungi. The form of silver used by physicians is known as Argentum metallicum.

Silver is an important trace mineral and minute particles were found in soil before it became depleted by overuse and now the only way to get the silver one needs is in liquid forms. According to several scientific journals published in the 20th century, scientists report that silver can actually double the amount of white blood cells in the body, our body's main germ fighters. The FDA has no evidence of harmful side effects and the US government's Centers for Disease Control and Prevention (CDC) confirms there has never been a reported allergic, toxic, or cancerous reaction in the use of liquid silver (1995). Excess silver can be deposited in the skin and tissues and may cause discoloration.

Honey and Sugar Heal Wounds

In normal situations, like a cut finger or a scraped knee, the wound will heal on its own. But sometimes wounds can become infected with bacteria which feed on the injured, causing more tissue damage, the wound starts to smell and pus collects. This pus is made up of dead white blood cells, part of the body's effort to kill bacteria and heal the infection.

Honey has four main properties that help to kill and to exterminate bacteria. There is lots of sugar in honey, in fact, it is a supersaturated solution of sugar, which means it contains as much sugar as can possibly be dissolved in it. This makes it taste great, but it also means it contains very little water. Applied to a wound, the honey acts like a dry sponge and soaks up any spare fluid. Because of this osmosis, the honey draws fluid away from the infected wound, this helps to kill bacteria, which need liquid to grow.

Honey is very acidic, with a pH between 3 and 4, about the same as orange juice or a can of coke.

Bacteria are killed in acidic environments like this. But if the honey is diluted (for example by the release of body fluids from the wound), it allows bacteria to grow again.

Honey produces hydrogen peroxide. Hydrogen peroxide is an antibacterial substance that is sold in pharmacies. It is made naturally in honey by an enzyme called glucose oxidase, which is added to the plant nectar by bees. Glucose oxidase is not active in full strength honey because of the honey's high acidity. However, when the honey is diluted (for example by the release of body fluids from the wound) the honey becomes less acidic, the enzymes become active, and hydrogen peroxide is produced.

Some honeys have antibacterial action that appears to be caused by phyto-chemicals found naturally in the nectar the bees collect. For example, honey made from the flowers of New Zealand's Manuka trees seems to be particularly powerful at killing bacteria. Because the exact molecule responsible for this activity has not yet been identified, it is called the Unique Manuka Factor.

Not only wounds are healed by honey, there are also anti-viral properties. Honey could also save limbs that otherwise need to be amputated, as our standard medical treatments fail. Honey seems capable of combating the growing surge of drug-resistant wound infections, especially methicillin-resistant Staphylococcus aureus (MRSA), or the infamous flesh-eating strain. These have become alarmingly more common in recent years, with MRSA alone responsible for half of all skin infections treated in the U.S. emergency rooms.

So-called super-bugs cause thousands of deaths and disfigurements every year, and public health officials are alarmed. Though the practice is uncommon in the United Stated, honey is successfully used elsewhere on wounds and burns that are unresponsive to other treatments. Attempts in the lab to induce a bacterial resistance to honey have failed possibly due to honeys complex attack, which might make adaption impossible. The more we keep giving antibiotics, the more we breed these super-bugs and wounds end up being repositories for them.

By eradicating them, honey could do a great job for society and to improve public health. Some of the most promising results come from Germany's Bonn University Children Hospital. Where doctors have used honey to treat wounds in 50 children whose normal healing processes were weakened by chemotherapy. The children fared consistently better than those with the usual applications of iodine, antibiotics and silver-coated dressings.

Sugar – everybody loves sugar, here is another reason to keep it in your cupboard. Sugar can heal cuts, scrapes, burns, and even large wounds without leaving a scar. It kills germs and repairs tissue better than any antiseptic or disinfectant on the market and may be the first known antiseptic in history. People have written about its miraculous properties for over 4000 years, since early Egyptian times, but it fell out of favor once antibiotics became available. Sugar and honey both contain high levels of glucose, the kind of sugar your body uses for energy and both are almost equal in their ability to heal, with honey taking a slight lead.

Sugar and honey prevent scarring to the extent that they heal ulcers and burns without the need of skin crafts. Scientists theorize both encourage the production of hyaluronic acid (HA), which fills out the skin by absorbing 3000 times its weight of water. At the same time, sugar and honey prevent the buildup of the stringy type of collagen that creates scar tissue. Instead it forms a different, mesh-like structure that brings the skin's surface back to normal and allows it to heal.

The interlinked history of microbes and humanity and identifying key changes in the way humans have lived, such as our move from hunter-gatherer to farmer to city-dweller, made us even more vulnerable to microbe attacks and our live styles today, with increased crowding and air travel, puts us once again at risk.

Chapter 16
DISEASES AND EPIDEMICS

Infectious Diseases

Bacteria existed long before humans evolved, and bacterial diseases probably co-evolved with each species which involuntarily hosts them. Many bacterial diseases that we see today have been around for as long as we have, others may have developed later. In either case, for the longest time we were not aware of the cause of infectious diseases. With the beginning of microbiology, bacterial pathogens became apparent.

Combining tales of devastating epidemics with accessible science and fascinating developments, it is interesting how closely microorganisms have evolved with us over the millennia, shaping human civilization through infection, disease, and deadly pandemics. Infectious disease caused by microbes are responsible for more deaths worldwide than any other single cause.

Unfortunately microbes are much better at adapting to new environments than people. Having existed on Earth for billions of years, microbes are constantly challenging humans with ingenious new survival tactics. Some infectious diseases such as the common cold, usually do not require a visit to your health care provider. However, some more serious conditions may require a physical examination, blood or urine tests and sometimes biopsies to identify certain bacteria and viruses.

How an infectious disease is treated depends on the microbe that cause it and sometimes on the age and medical condition of the person infected. Certain diseases are not treated at all, but are allowed to run its course, with the immune system doing its job alone.

Some diseases, such as the common cold, are treated only to relieve the symptoms. Others, such as strep throat, are treated to destroy the offending microbe as well as to relieve symptoms. The immune system has an arsenal of ways to fight off invading microbes. Most begin with B and T-cells and antibodies whose sole purpose is to keep the body healthy. Some of these cells sacrifice their lives to rid us of disease and restore our body to a healthy state.

Other important ways to react to an infection may include: Inflammation, vomiting, diarrhea, fatigue, cramping, mucous production, fever, coughing and sneezing. Some disease causing microbes can make you very sick quickly and then not bother you again. Some can last for a long time and continue to damage tissue. Others can last forever, but you will not feel sick anymore, or you will feel sick only once in a while. Most infections cause by microbes fall into three major groups.

***Acute infections** – are usually severe and last a short time. They can make you feel very uncomfortable, with signs and symptoms such as tiredness, achiness, coughing, and sneezing. The common cold is such an infection and may last from two to twenty-four days.

Chronic infections – usually develop from acute infections and can last for days to months to a lifetime. Sometimes people are unaware they are infected but still may be able to transmit the germ to others.

Latent infections – are hidden or silent and may or may not cause symptoms again after the first acute episode. Some infectious microbes, usually viruses, can wake up and become active again but not always causing symptoms. With the arrival of antibiotics and modern vaccines, as well as important sanitation and hygiene, many diseases that formerly posed an urgent threat to public health were brought under control or largely eliminated. As a result, by the mid-20th century some scientists thought that medical science had conquered infectious diseases.

Despite these public health advances, new microbes emerge and old microbes re-emerge just as they have throughout history. Several pressures contribute to the emergence of diseases, such as: redistribution of human populations, rapid global travel, changes in the way we use land and ecological, environmental and technical changes.

Practices such as misuse of antibiotic medicines also contributes to disease emergence. In addition unsanitary conditions in animal agriculture and increasing commerce in exotic animals (for food or pets) have contributed to the rise in opportunity for animal microbes to transfer from animals to humans.

The reappearance of microbes that had been conquered successfully or controlled by medicines and vaccines is distressing to the scientific and medical communities, as well as to the public. One major cause of disease re-emergence is that many microbes responsible for causing these diseases are becoming resistant to the drugs used to treat them.

Some examples of re-emerging infectious diseases that are of significant public health concern are dengue, malaria, tuberculosis and polio. Scientists usually define emerging microbes as those that have appeared only recently in a population or have existed but are rapidly in incidence or geographic range.

Recent examples of such disease-causing microbes are methicillin-resistant Staphylococcus aureus (MRSA) bacteria, West Nile virus, and 2009 H1N1 influenza virus. What do we know of the bacterial diseases that bothered our ancestors? Quite a few infectious diseases are known from historical times. Probably the best known is the Black Death or Plague, caused by yersinia pestis and Tuberculosis were major killers in Europe in the past.

Epidemics raged over countries, followed trade routes and travelers or colonists, and could wipe out a considerable proportion of inhabitants. The results were often devastating, and treatment was limited due to lack of knowledge. The big killers in the Americas were European diseases, often caused by viruses. What kindled such epidemics? That depends on the causative agent. Human-to-human contact is needed for many viruses to spread, (Ebola or the common cold), fecal-oral contamination (most frequently drinking water contaminated with sewage) causes most enteric disease outbreaks including cholera, whereas insects (fleas or mosquitos) may be needed for the spread of yet other diseases.

Contrary to common believe, corpses are not a common source of infection. Most post-catastrophic outbreaks are the result of lack of clean drinking water. In some cases we can identify with various degrees of certainty which infection caused the death or suffering of famous people. "Pharaoh I" may have suffered from ear infection (Source: CNN) and smallpox may have brought "Ramses V", "Queen Mary II" of England, and "Louis XV" of France together. "Alexander the Great" died of an infection of the lungs, possibly causes include West Nile Virus; before him, his dear friend "Hephaistos" probably died of typhoid fever, according to the symptoms described of his death bed. "Amadeus Mozart" may have died of rheumatic fever, which is caused by a prior infection. Several famous people suffered from syphilis, including the composer "Franz Schubert", the author "Karen Blixen" and the painter "Vincent van Gogh", the disease could be lethal in pre-antibiotic days.

Clearly, epidemics are not something of the past. We have new and old diseases to combat. But those massive killings known from historian times are no longer common in most parts of the world. This was mainly achieved by vaccination after the organisms causing the disease were identified and characterized and vaccines developed.

Some of the Worst Pandemics in History

Smallpox – 430 BC? to 1979. Smallpox is a contagious disease unique to humans and is caused by either of two virus variants named Variola major and Variola minor. The deadlier form V.major, has a mortality rate of 30-35%, while V.minor causes a milder form of disease called alastrim and kills 1% of its victims. Long-term side effects for survivors include the characteristic skin scars. Occasional side effects include blindness due to the corneal ulcerations and infertility in male survivors.

Smallpox killed an estimated 60 million Europeans in the 18th century alone. Up to 30% of those infected, including 80% of the children under 5 years of age, died from the disease, and one third of the survivors became blind. As for the Americas, after the first contacts with Europeans and Africans, some believe that the death of 90-95 percent of the native population of the New World was caused by Old World diseases. It is suspected that smallpox was the chief culprit and responsible for killing nearly all of the native inhabitants of the Americas. Smallpox was responsible for an estimated 300-500 million deaths in the 20th century. As recently as 1967, the World Health Organization (WHO) estimated that 15 million people contracted the disease and that two million died in that year. After successful vaccination campaigns throughout the 19th and 20th centuries, the WHO certified the eradication of smallpox in 1979. To this day, smallpox is the only human infectious disease to have been completely eradicated from nature.

Spanish Flu – 1918 to 1919. Killed 50 to 100 million

people worldwide in less than two years. In 1918 and 1919 the Spanish Flu pandemic killed more people than Hitler, nuclear weapons and all the terrorists of history combined. Spanish Influenza was a more severe version of the typical flu, with the usual sore throat, headaches and fever. However, in many individuals, the disease quickly progressed to something much worse than sniffles. Extreme chills and fatigue were often accompanied by fluid in the lungs.

One doctor treating the infected described the grim scene. The faces were a bluish cast, a cough brings up the blood-stained sputum and in the morning, the dead bodies are stacked about the morgue like cordwood. If the flu passed the stage of being a minor inconvenience, the patient was usually doomed. There is no cure for the influenza virus, even today. All doctors could do was try to make the patients comfortable, which was difficult with their lungs filled with fluid as they were wracked with unbearable coughing. The bluish cast on their faces eventually turned brown or purple and their feet turned black. The lucky ones simply drowned in their own lungs. The unlucky ones developed bacterial pneumonia as an agonizing secondary infection. Since antibiotics had not been invented yet, this too was essentially not treatable.

The pandemic came and went like a flash. Between the speed of the outbreak and military censorship of the news during World War I, hardly anyone in the United States knew that a quarter of the nation's population – and a billion people worldwide – had been infected with the deadly disease. More than half a million people died in the United States alone, and worldwide more than 50 million.

***_Black Death_** – 1340 to 1771, or Black Plague, was one of the deadly pandemics in human history. It began in South-western or Central Asia and spread to Europe by the late 1340s. The total number of deaths worldwide from the pandemic is estimated at 75 million people. There were an estimated 20 million deaths in Europe alone and the Black Death is estimated to have killed between a third and two-thirds of Europe's population.

The three forms of plague brought an array of signs and symptoms to those infected. Bubonic plague refers to the painful lymph node swelling called buboes, mostly found around the base of the neck, and in the armpits and groin. The septicaemic plague is a form of blood poisoning, the pneumonic plague is an airborne plague that attacks the lungs before the rest of the body. The classic sign of bubonic plague was the appearance of buboes in the groin, neck and armpits, which oozed pus and bled. Victims underwent damage to the skin and underlying tissue, until they were covered in dark blotches.

Most victims died within four to seven days after infection. When the plague reached Europe, it first struck port cities and then followed the trade routes, both by land and sea. The bubonic plague was the most commonly seen form during the Black Death, with a mortality rate of thirty to seventy-five percent and symptoms including fever of 38-41C (101-105F), headaches, painful aching joints, nausea and vomiting, and a general feeling of malaise. Of those who contracted the bubonic plague, 4 out of 5 died within eight days. Pneumonic plague was the second most commonly seen form with a mortality rate of ninety to ninety-five percent.

The same disease is thought to have returned to Europe every generation with varying virulence and mortalities until the 1700s. During this period, more than 100 plague epidemics swept across Europe. On its return in 1603, the plague killed 38,000 Londoners. Other notable 17th century outbreaks were the Italian Plague of 1629-1631, the Great Plague of Seville (1647-1652), the Great Plague of London (1665-1666) and the Great Plague of Vienna (1679).

There is some controversy over the identity of the disease, but in its virulent form, after the Great Plague of Marseille in 1720 1722 and the 1771 plague in Moscow it seems to have disappeared from Europe in the 18th century. The fourteenth-century eruption of the Black Death had a drastic effect on Europe's population, irrevocably changing Europe's social structure. It was a serious blow to the Roman Catholic Church and resulted in widespread persecution of minorities such as Jews, foreigners, beggars and lepers. The uncertainty of daily survival created a general mood of morbidity influencing people to live for the moment

***Malaria** – 1600 to today. Malaria cause about 400-900 million cases of fever and approximately one to three million deaths annually, this presents at least one death every 30 seconds. The vast majority of cases occur in children under the age of 5 years, pregnant women are also especially vulnerable. Despite efforts to reduce transmission and increase treatment, there has been little change in which areas are at risk of this disease since 1992.

Indeed, if the prevalence of malaria stays on its present upwards course, the death rate could double in the next twenty years. Precise statistics are unknown because many cases occur in rural areas where people do not have access to hospitals or the means to afford health care. Consequently, the majority of cases are undocumented.

Malaria is one of the most common infectious diseases and an enormous public-health problem. Its parasites are transmitted by female Anopheles mosquitoes. The parasites multiply within red blood cells, causing symptoms that include symptoms of anemia (light headedness, shortness of breath, tachycardia, etc.) as well as other general symptoms such as fever, chills, nausea, flu-like illness, and in severe cases, coma and death. The disease is caused by protozoan parasites of the genus Plasmodium. It is widespread in tropical and subtropical regions, including parts of the Americas, Asia and Africa.

*_**Aids**_ (1981 to present) Acquired Immune Deficiency Syndrome (AIDS) has led to the deaths of more than 25 million people since it was first recognized in 1981, making it one of the most destructive epidemics in recorded history. Despite recent improves access to antiretroviral treatment and care in many regions of the world, the AIDS epidemic claimed approximately 3.1 million (between 2.8-3.6 million) lives in 2005 (an average of 8,500 per day), of which 570,000 were children. UNAIDS and the

WHO estimate that the total number of people living with the human immunodeficiency virus (HIV) has reached its highest level. There are an estimated 40.3 million (estimated range between and 45.3 million) people now living with HIV. Moreover, almost 5 million people have been estimated to have been infected with HIV in 2005 alone.

The pandemic is not homogenous within regions with some countries more afflicted than others. Even at the country level there are wide variations in infection levels between areas. The number of people living with HIV continues to rise in most parts of the world, despite strenuous prevention strategies. Sub-Saharan Africa remains by far the worst-affected region, with 23.8 million to 28.9 million people living with HIV at the end of 2005, 1 million more than in 2003. Sixty-four percent of all people living with HIV are in Sub-Saharan Africa, as are more than 77% of all women living with HIV. South and South East Asia are second most affected with 15%.The key facts surrounding this origin of AIDS are currently unknown, particularly where and when the pandemic began, though it is said that it originated from the apes in Africa.

**Cholera* (1817 to present) – 8 epidemics, hundreds of thousands killed worldwide. In the 19th century, Cholera became the world's first truly global disease in a series of epidemics that proves to be a watershed for the history of plumbing. Festering along the Ganges River in India for centuries, the disease broke out in Calcutta in 1817 with grand-scale results. When the festival was over, they carried cholera back to their homes in other parts of India. There is no reliable evidence of how many Indians perished during that epidemic, but the British army counted 10,000 fatalities among its imperial troops.

Based on those numbers, it is almost certain that at least hundreds of thousands of natives must have fallen victim across the land. Cholera sailed from port to port, the germ making headway in contaminated kegs of water or in the excrement of infected victims, and transmitted by travelers. The world was getting smaller thanks to steam-powered trains and ships, but living conditions were slow to improve. By 1827 cholera had become the most feared disease of the century. The major cholera pandemics are generally listed as: First: 1817-1823, Second: 1829-1851, Third: 1852-1859, Fourth: 1863-1879, Fifth: 1881-1896, Sixth: 1899-1923, Seventh: 1961-1870, and some would argue that we are in the Eight: 1991 to the present. Each pandemic, save the last, was accompanied by many thousands of deaths. As recently as 1947 – 20,500 of 30,000 people infected in Egypt died. Despite modern medicine, cholera remains an efficient killer.

**Typhus* (430 BC? To present) – Killed 3 million people between the years 1918-1922 alone, most of Napoleon's soldiers in Russia. Typhus is one of several similar diseases caused by louse-home bacteria. The name comes from the Greek typhus (smoky or lazy), describing the state of mind of those infected with typhus. Rickettsia is endemic in rodent hosts, including mice and rats, and spreads to humans through mites, fleas and body lice. The arthropod vector flourishes under conditions of poor hygiene, such as those found in prisons or refugee camps, the homeless, or until the middle of the 20th century, in armies in the field.

The first description of typhus was probably given in 1083 at a convent near Salerno, Italy. Before a vaccine was developed in World War II, typhus was a devastating disease for humans and has been responsible for a number of epidemics throughout history. During the second year of the Peloponnesian war (430 BC), the city-state of Athens in ancient Greece was hit be a devastating epidemic, known as the Plague of Athens. The plague returned twice more, in 429 BC and in the winter of 427/6 BC. Epidemic typhus is one of the strongest candidates for the cause of this disease outbreak, supported by both medical and scholarly opinions. Epidemics occurred throughout Europe from the 16th to the 19th centuries and during the English Civil War, the Thirty Years War and the Napoleonic Wars. During Napoleon's retreat from Moscow in 1812, more French soldiers died of typhus than were by the Russians. A major epidemic occurred in Ireland between the years 1816-1819, and again in the late 1830s, and yet another major typhus epidemic occurred during the Great Irish Famine between 1846 and 1849.

In America a typhus epidemic occurred in Hampshire in 1843 and struck Philadelphia in 1837. Several epidemics occurred in Baltimore, Memphis and Washington DC between 1865 and 1873. During World War 1 typhus caused three million deaths in Russia and more in Poland and Romania.

De-lousing stations were established for troops on the Western front but the disease ravaged the armies of the Eastern front, with over 150,000 dying in Serbia alone. Fatalities were generally between 10 to 40 percent of those infected and the disease was a major cause of death for those nursing the sick. Following the development of antibiotics during World War II epidemics now occur only in Eastern Europe, the Middle East and parts of Africa.

Chapter 17
QUARANTINE AND ISOLATION

40 Days of Separation

Quarantine – often a compulsory isolation, typically to contain the spread of something considered dangerous, often, but not always, disease. The word comes from the Italian – quarantena, meaning forty-day period.

The practice of quarantine – the separation of the diseased from the healthy – has been around for a long time. As early as the writing of the Old Testament, for instance, rules existed for isolating lepers. It was not until the Black Death of the 14th century, however, that Venice established the first formal system of quarantine, requiring ships to lay at anchor for 40 days before landing. The Venetian model was practiced until the discovery in the late 1800s that germs cause disease, after which health officials began tailoring quarantines with individual microbes in mind.

In the mid-20th century, the advent of antibiotics and routine vaccinations made large-scale quarantines a thing of the past, but today bio-terrorism and newly emerged diseases like SARS threaten to resurrect the age-old custom, potentially on the scale of entire cities. Quarantine periods can be very short, such as in the case of a suspected anthrax attack, in which persons are allowed to leave as soon as they shed their potentially contaminated garments and undergo a contamination shower. The purpose for such de-contamination is to prevent the spread of disease and to contain contamination so others are not put at risk.

People can be infected with dangerous diseases in a number of ways. Some germs, like those causing malaria, are passed to humans by animals. Other germs, like those that cause botulism, are carried to people by contaminated food or water. Still others, like the ones causing measles, are passed directly from person to person.

These diseases are called contagious. State and local health departments have created emergency preparedness and response plans to implement early detection, rapid diagnosis and treatment with antibiotics and antivirals. These plans use two main traditional strategies – quarantine and isolation – to contain and to control the spread of contagious disease by limiting people's exposure to it.

1) Quarantine applies to those who have been exposed to a contagious disease but may or may not become ill.

2) Isolation applies to persons who are known to be ill with a contagious disease.

Isolation and Special Care

When someone is known to be infected with a contagious disease, they are placed in isolation and receive special care, with precautions taken to protect the uninfected.

Historically speaking, people with the bad luck to develop an infection have never had it so good. Modern medicine can deploy a vast array of antibiotics and other remedies for their benefit. For some of them, though, it includes a direct carry-over from the "Middle Ages". These are the people who are not just infected on the inside but also infested on the outside, covered with germs. When they are hospitalized we hustle them into an isolation room, and no matter how much they protest and complain, and no matter how cumbersome it makes the rest of their medical care, we do not let them out.

We keep patients with tuberculosis in rooms specially ventilated, so that in theory, germs do not rush out into the public corridor when the door is opened. All visitors wear tight-fitting masks, but gloves and gowns are unnecessary, as tuberculosis does not spread by touch.

Touch does transmit methicillin-resistant staph, or MRSA, and other antibiotic-resistant bacteria that are considered a problem in many hospitals these days. Some of the isolation rooms are occupied by patients harboring these germs, but most are in use by patients with the intestinal infection called C.difficile colitis.

However, both quarantine and isolation may be conducted voluntarily or compelled on a mandatory basis through legal authority, e.g. the first astronauts to visit the moon were quarantined upon their return at the specially designed Lunar Receiving Laboratory.

Chapter 18
PREVENTION OF DISEASE

Vaccination – Inoculation

Vaccination is the administration of antigenic material (vaccine) to stimulate adaptive immunity to a disease. Vaccines can prevent or ameliorate the effects of infection by many pathogens.

Inoculation is the placement of something that will grow or reproduce, and is most commonly used in respect of the introduction of a serum, vaccine, or antigenic substance into the body of a human or animal, especially to produce or boost immunity to a specific disease. It can also be used to refer to the communication of a disease to a living organism by transferring its causative agent into such.

Today the terms inoculation, vaccination and immunization are used more or less interchangeably and popularly refer to the process of artificial induction of immunity against various infectious diseases. The micro-organism used in an inoculation is called inoculant or inoculum.

Some form of microbiological treatment was practiced by many people, throughout the two hundred year period after Hooke and van Leeuwenhoek, but this was with very little understanding of the microbial processes involved.

Primitive forms of smallpox inoculation developed in Turkey in the seventeenth century were brought to England around 1720. This involved creating reasonably large open wounds in the arm with a knife and pasting in serum taken from the pus of victim's sores. It was effective – sometimes – and even some members of the royal family underwent the procedures. Elsewhere in England, cowherds and milkmaids had noticed the immunizing effects of cowpox in apparently preventing smallpox, and they too practiced a form of open-wound inoculation on local people.

However, the credit for the invention of inoculation goes to Edward Jenner who in 1796 injected cowhand James Phipps with cowpox. Jenner gets the credit because he carried out his work using accepted scientific methods and wrote it up afterwards, though the ethics of deliberately injecting the experimental subject with smallpox is questionable!

Fortunately, the boy did not develop smallpox and Jenner became rich and famous as a result. It was from this risky beginning that the science of immunization developed.

Smallpox (*Variola virus*) was declared officially eliminated in 1979 – the only microbial disease ever deliberately eradicated. While many scientists of the eighteenth and nineteenth century studied plant and animal structures under the microscope, the real science of microbiology only began in the latter half of the nineteenth century, when high-magnification microscopes of good optical quality became more widely available. The most notable person was Ferdinand J. Cohn who in 1875 effectively founded the science of bacteriology (a branch of microbiology which studies bacteria). His main contribution was the classification of bacteria, he named the Bacillus.

Immunization

When germs such as viruses or bacteria invade the human body, the immune system makes special cells. These cells produce antibodies, which help destroy these germs. The next time the body is exposed to the same infection, the immune system often recognizes it and sets out to destroy it. Immunizations work in much the same way. The exposure is to a very small, very safe amount of a virus or bacteria that has been weakened or killed. The immune system then learns to recognize and attack the infection if there is exposure to it later in life. As a result, one will either not become ill or have a milder infection. This is a natural way to deal with infectious diseases.

After immunizations were introduced on a wide scale, infections such as tetanus, diphtheria, mumps, measles, pertussis (whooping cough), and polio became rare. Newer immunizations have also decreased certain types of meningitis, pneumonia, and ear infections in children.

Four different types of vaccines are currently available.

Attenuated (weakened) live virus – is used in the measles, mumps, rubella (MMR) and the varicella (chickenpox) vaccine.

Killed (inactivated) vaccines – are made from a protein or other small pieces taken from a virus or bacteria. These vaccines are safe, even in people with weakened immune systems. Influenza shots are an example of this type of vaccine.

Toxoid vaccines – contain a toxin or chemical made by the bacteria or virus. They make you immune to the harmful effects of the infection, rather than to the infection itself. Examples are the diphtheria and tetanus vaccines.

Biosynthetic vaccines – contain human-made substances that the immune system thinks are infectious organisms.

Most vaccines are given by hypodermic injection as they are not absorbed reliably through the intestines. Live attenuated polio, some typhoid and some cholera vaccines are given orally in order to produce immunity based in the bowel.

Chapter 19

CHEMICALS–PATHOGENS–VIRUSES

Biological Warfare

Jiang Chun Geng's poisoned right leg, with its suppurating wounds, hangs limply over the gray wooden bench in the medical clinic in Dachen, a village in China's province of Zhejiang. Twice the size of his left leg, the limb is tender to touch, while the doctor gently dabs the putrid openings with an alcohol-drenched swab. Jiang's heavily lined face tightens as the doctor wraps the fiery stump with a white bandage and unhooks an intravenous antibiotic drip. Another treatment is over. He cannot remember a time when flesh-eating ulcers did not cover his legs, the wounds first appeared in July 1942 after the Japanese army passed through the village. His entire family developed the festering sores and his mother and younger brother died in unbearable pain a decade later, as the untreated, mysterious infection crept up their legs.

Biological or bio-warfare is the use of biological pathogens (bacteria, viruses, fungi, and toxins) derived from living organisms to kill or incapacitate one's enemies. From poisoned arrows (Scythians, and later the Viet Cong guerillas) to poisoned wells (Sparta, Persia, Rome and others) to bombs with deadly bacteria (Japan, United States, Soviet Union and Iraq), the intentional use of bio-warfare has been around for centuries.

Biological warfare also known as germ warfare is the deliberate use of disease-causing biological agents to kill or incapacitate humans, animals or plants. Biological weapons (often referred to as bio-weapons) are living organisms or replicating entities (virus) that reproduce or replicate within their host victims.

Biological weapons may be employed in various ways to gain strategic or tactical advantages over an adversary, either by threat or by actual deployment. Like some of the chemical weapons, biological weapons may also be useful as area denial weapons. These agents may be lethal or non-lethal, and may be targeted against a single individual, or a group of people, or even an entire population. They may be developed, acquired, stockpiled or deployed by states or non-national groups.

There is an overlap between biological and chemical warfare, as the use of toxins produced by living organisms is considered under the provisions of both the Biological Weapons Convention and the Chemical Weapons Convention. Toxins and psycho-chemical weapons are often referred to as mid-spectrum agents.

Unlike bio-weapons, these mid-spectrum agents do not reproduce in their host and are typically characterized by shorter incubation periods. Ideal characteristics of a biological agent to be used as a weapon against humans are: highly effective, high virulence, non-availability of vaccines, and effective and efficient delivery system. Stability of the weaponized agent (ability of the agent to retain its infectivity and virulence after a prolonged period of storage) may also be desirable, particularly for military applications.

Biological warfare can specifically target plants to destroy crops or defoliate vegetation and attacking animals is another area of biological warfare intended to eliminate animal resources for transportation and food.

It is important to note that all of the classical and modern biological weapons organisms are animal diseases, the only exception being smallpox. Thus, in any use of biological weapons, it is highly likely that animals will become ill either simultaneously with, or perhaps earlier than humans.

Microbes have played a profound role in warfare, religion, migration of populations, art and diplomacy. Boundaries of nations have changed as a result of microbial diseases and infectious agents have terminated some kingdoms and elevated others.

The traditional approach toward protecting agriculture, food and water, focusing on the natural or unintentional introduction of a disease is being strengthened by focused efforts to address current and anticipated future biological weapons threats that may be deliberate, multiple and repetitive.

Biological weapons allow for the potential to create a level of destruction and loss of life far in excess of nuclear, chemical or conventional weapons, relative to their mass, cost of development and storage. Germs and other micro-organisms are considered to be our enemies and we do our best to destroy them. So far, we have not succeeded. But what would happen if we did?

A World without Bacteria

A world without bacteria may seem ideal at first. But the more we think about it, the more we realize that we cannot live without them. There would be no food poisoning, no diarrhea, no coughs and colds, no sore throats, no tuberculosis, no cholera, no small pox, no polio, no sexually transmitted diseases . . . There would be no decay. Foods would not get spoiled and we would not need preservatives, refrigeration and wasteful packaging.

No decay – what happens then to all the plants, animals and people at the end of their life cycles? They would be preserved forever and this whole world would be filled with dead bodies piled up over millions of years. We would be living in a giant graveyard. We cannot live in a world without decay, there would not be enough room for the living.

Human Decomposition

Decomposition is the process by which organic material is broken down into simpler forms of matter. Five general stages are used to describe this process: Fresh, Bloat, Active and Advanced Decay, and Dry/Remains. The general stages of decomposition are coupled with two stages of chemical decomposition: autolysis and putrefaction, which break down the main components of the human body.

The University of Tennessee's Medical Center has a lovely forest grove with squirrels in the trees, birds calling and patches of green grass where bodies lie on their backs in the sun, or sometimes in the shade, depending on where the researchers put them.

> *This pleasant Knoxville hillside is a field research facility, the only one in the world dedicated to the study of the human decay. The people lying in the sun are dead. They are donated cadavers, helping, in their mute, fragrant way, to advance the science of criminal forensics. The knowledge gained about a dead bodies decay - the biological and chemical phases they go through, how long each phase lasts, how the environment affects these phases - the better equipped we are to figure out when a given body died.*

To understand how these variables affect the time of decomposition, one must be intimately acquainted with the control scenario: basic unadulterated human decay. When we let nature take its course, just exactly what course does it take?

The hallmark of fresh-stage decay is a process called autolysis, or self-digestion. Human cells use enzymes to cleave molecules, breaking compounds down into usable parts. When a body dies, all the immune responses start to quickly turn off, and general cell necrosis sets in as bacteria start to invade the dying and dead cells (most cells die quickly from lack of oxygen).

While a person is alive, their cells keep the enzymes in check, preventing them from breaking down cell's own walls. After death, the enzymes operate unchecked and begin eating through the cell structure, allowing the liquid inside to leak out, making its way through the body to allow contact with the body's bacteria colonies to assist in putrefaction. They have already been enjoying the benefits of the decommissioned immune system, and now, suddenly, they are presented with this edible goo from the raptured cells of the intestine lining. As will happen in times of abundant food supply, their population swells. Soon bacteria are everywhere and the scene is set for stage two: Bloat.

The life of bacteria is built around food, they do not have mouths or fingers, but they eat, they digest, they excrete. Like us, they break their food down into its more elemental components. When we die, they stop feeding on what we have eaten, they begin feeding on us and produce gas in the process. Intestinal gas is a waste product of bacteria metabolism. Bloating is most noticeable in the abdomen, where the largest number of bacteria are, is typically short lived, perhaps a week and the final stage, putrefaction and decay, last the longest. The end of decay is signaled by the migration of maggots away from the body. Microbes exist to bring back to the Earth that which is weakened.

Orthodox Jews shroud their dead and bury them on the same day as death, while Buddhists believe that consciousness stays in the body for three days. Hindus are cremated, because it is believed that burning releases the soul from the body, while Roman

Catholics frown on cremation out of respect for the body as a symbol of human life.

We need decay to break down dead matter into basic substances to provide nutrients for new plants to grow. Otherwise, there would not be any food for animals and humans to survive. We too, need decay to break our food down so that nutrients can be absorbed by the body. We tend to think that digestion takes place in the stomach, with the help of acids. That is only a small part of the process, since food stays in the stomach for only three to five hours and digestion continues in the intestines where food is broken down (decayed) by bacteria.

We develop powerful antibiotics that interfere with our healthy gut flora – we kill them with chemicals, radiation and other means when disease strikes. Many contributions by scientists and researchers have been made, not always accepted by the medical community.

Chapter 20
RESEARCH

Royal Rife – Successfully Destroyed Cancer

Royal Rife – a pioneer in Energetic Medicine Research identified the human cancer virus, using the world's most powerful microscope (which Rife created), cultured the virus (on salted pork, the best growing medium), and injected it into rats. These injections caused cancer in every one of the rats (all 400 of them). Later Rife was able to find a frequency of electromagnetic energy that would cause the cancer virus to self-destruct when within that energy field. Rife created a device that emitted that energy field and was successful at destroying cancer viruses inside patients who were within close proximity of this device.

In 1934, the University of Southern California appointed a Special Medical Research Committee which brought 16 terminal cancer patients from Pasadena County Hospital to Rife's laboratory and clinic for treatment. This team included doctors and pathologists assigned to examine the patients, if still alive, in 90 days. After three months of treatment, the Committee concluded that 14 of the patients had been completely cured and the remaining two were cured within the following four weeks. On the evening of a press conference to announce the results of the 1934 study on Rife's cancer therapy, the former president of Southern California AMA, was fatally poisoned and his papers were lost. Shortly after Rife's labs were destroyed by arson and sabotage and Royal Rife himself was killed in 1971.

By the late 1950's even independent doctors and labs (among them New York's Presbyterian Hospital) – unaware of Rife's work, were bringing forth their own proof that cancer was an infectious viral disease, and maintained that cancer did not consist of a localized tumor alone . They described it as a generalized disease caused by an organism in the blood.

It could be that all of us have that virus and that it does not create tumors except when there is enough toxicity due to chemical exposure, unhealthy habits, or poorly oxygenated blood. **Microbes exist to bring back to the Earth that which is weakened.**

Royal Rife was a brilliant scientist – he began his career as a research pathologist and was a medical researcher of the highest qualifications. In the early 1930's he invented the first, and currently still the most powerful universal microscope which had a magnification of 60.000 diameters and a resolution of 31.000 diameters. Dr. Rife was the first person to see living viruses since the most powerful electron microscopes today will kill the organism being observed. He would work sometimes as long as twenty hours to get a virus into focus to allow him to observe it in its changing forms, thus providing evidence that microbes and viruses have the ability to change from one form to another depending upon the medium they are in.

Since the days of Dr. Nikola Tesla, men of science had known of the connection to electrical, magnetic and radio frequencies within the operation of the human body, so Rife turned to electro-magnetic frequencies since he could expose the bacteria and virus to such frequencies and watch the effects. He discovered that each microbe and each virus had a particular frequency to which it was vulnerable. Rife called this the mortal oscillatory rate, a term still used to this day.

In the nineteenth century, Semmelweiss struggled to convince surgeons that it was a good idea to sterilize their instruments and use sterile surgical procedures. Pasteur was ridiculed for years for his theory that germs could cause disease. Scores of other medical visionaries experienced fierce opposition for simply challenging the medical status quo of the day, including legends such as Roentgen and his X-rays, Morton for promoting the absurd idea of anaesthesia, Harvey for his theory of the circulation of the blood, and many others in recent decades including: W.F.Koch, Revici, Burzynski, Naessens, Priore, Hamer, Clarke, Livingstone-Wheeler, Reich, and Hoxsey.

Hamer – New German Medicine

In 1981, Dr. Hamer, a German doctor, discovered a revolutionary Brain-Emotion Disease connection, summarized his research with this statement: I searched for cancer in the cell and I have found it in the form of a wrong coding in the brain.

The following articles will give an introduction to Dr. Hamer's "German New Medicine", which some consider the current cutting-edge in healing arts. From his personal experience – Dr. Hamer himself had cancer – and from those patients he took care of in many university clinics, he established over the years that there is always a definite syndrome at the source of cancer, and not just some kind of stress. It requires a strong stimulus, a brutal psychic trauma, which hits the patient as a major event in his life, an acute dramatic conflict and complete psychic isolation. This initial syndrome, which he discovered and carefully verified in each of the ca. 15,000 cases he has examined until now, he called Dirk Hamer Syndrome (DHS), after his son's name Dirk, whose tragic death in 1978 originated Dr. Hamer's own cancer.

Dr. Hamer himself developed testicular cancer soon after his son was shot dead in 1978, and wondered whether his son's death was the cause of his cancer. However, it wasn't until three years later, when he worked in a cancer hospital at the University of Munich, that Dr. Hamer was able to start his cancer research in earnest.

The experience of those thousands of individual cases diagnosed and treated during those years, allowed him to bring out constants and to formulate the **iron law of cancer,** which has never been contradicted. This law, based on the DHS is the main substance, the keystone and states as follows:

1) Any cancer starts with a DHS, that is to say, an extremely brutal shock, a dramatic and

acute conflict, experienced in loneliness and sensed by the patient as the most serious he has ever known.

2) It is the subjective meaning of the conflict, the way the patient experienced it at the moment of the DHS, which determines:

   a) The focus – that is, the specific area of the brain, which, under the influence of the psychic trauma, suffers a breakdown and thus produces an ill-proliferation of cells (cancer) in the organ dependent on this short circuited cerebral area;

   b) The location of the cancer in the organism;

   c) There is an exact correlation between the evolution of the conflict and the evolution of the cancer (tumor or hypo-function) on a double level: cerebral and organic.

A second or third conflict with DHS may follow in the wake of the first. For instance the diagnosis of cancer can create a sudden fear of death, with spots in the lungs, or a conflict of self-devaluation, with cancer of the bones: this is not metastasis but new tumors caused by new foci in the brain, occurring under the influence of new psychic trauma.

As soon as the conflict is resolved, there is an inversion of polarity, the breakdown in the brain is repaired by forming an oedema, whereas the cells whose anarchic proliferation was due to a coding error of the brain's computation, are no longer enervated and the tumor stops spreading, heals, is encysted, transformed or ejected.

Our body now thoroughly eliminates useless tumor cells or necrotic ulcer cells, as well as other wastes it harbors, then it generates new, fresh tissue wherever needed and rejuvenates. This is accomplished with the help from its allies, the microbes, which symbiotically help us clean out wastes from within.

The different microbes, the fungi, the bacteria and the viruses – some intrinsic, some designed to be added from outside – all have an affinity for the tissues derived from the same ontogenetic/phylogenetic origin.

The most ancient microbes, the fungi and mycobacteria, are related to all organs that developed out of the archaic endoderm and are controlled by the old brain stem. The **endoderm** germ layer contributes to the respiratory and gastrointestinal tracts and to all of their associated organs.

The less ancient ones, the bacteria, are related to the organs out of the mesoderm and are controlled by the cerebellum and the cerebral medulla. The **mesoderm** is the middle layer and relates to a number of tissues and structures including bone, cartilage, muscle, connective tissue (including that of the dermis), blood vascular, reproductive, excretory and urinogenital systems and contributes to some glands.

The more recent one, the viruses, are controlled solely with the organs that developed out of the ectoderm, controlled by the evolutionary newest brain formation, the cerebral cortex. The **ectoderm** the outermost of the three primitive germ layers of the embryo relate to the epidermis and form tissues, such as the nails, hair, and glands of the skin, the nervous system, external sense organs and mucous membrane of the mouth and anus.

The older brain parts steer organs to break down useless tumors and other waste with the help of fungi, mycobacteria and certain bacteria, while the newer brain parts steer organs to fill with the help of bacteria and viruses, preceded by bacterial cleansing of necrotic waste.

The claim by conventional physicians that viruses cause a number of cancer is correct, but it is our body which uses them for its different functions, in order to optimize health.

Dr. Hamer states that most people with cancer, even with mild pain, are treated with morphine in some form, which changes the brain, paralyzes the intestines and completely disables the restoration functions just when the patient was in the recuperation phase on his way to health. The pain is actually a very good sign of recovery, but modern medicine interprets it as the opposite, a sign of impending death. These substances are administered in order to "make the end easier" and sure enough, the person soon dies, but it is not of the cancer, it may be of the medication . . .

Emotion and Memory

Emotion can have a powerful impact on memory. Numerous studies have shown that most vivid autobiographical memories tend to be of emotional events, which are likely to be recalled more often and with more clarity and detail than neutral events.

The activity of emotionally enhanced memory retention can be linked to human evolution. During early development, responsive behavior to environmental events would have progressed as a process of trial and error. Survival depended on behavioral patterns that were repeated or reinforced through life and death situations. Through evolution this learning became genetically embedded in humans and all animal species in what is known as fight or flight instinct.

Artificially inducing this instinct through traumatic physical or emotionally stimuli essentially creates the same physiological condition that heightens memory retention by exciting neuro-chemical activity affecting areas of the brain responsible for encoding and recalling memory. This memory-enhancing effect of emotion has been demonstrated in a large number of studies, using stimuli ranging from words to pictures to narrated slide shows, as well as autobiographical memory studies. By re-creating past traumatic events out of memory the continuous cycle of emotional engagement is accomplished.

Chapter 21

CONSCIOUSNESS AND AWARENESS

Consciousness – The Binding Force

Humans realize that we are united with all living things. Nature does not recognize higher or lower levels of life – it orchestrates diversity in the eco-system. Vitality is the breath that sustains life by rhythmically joining us to Nature. We inhale all that is necessary to remain alive, then exhale it back to wherever it is needed next.

At birth most infants are very much alike physiologically, but by age seventy no two people's bodies are remotely the same

Time has made each of us unique. This material fact underlies a great deal of the separation in the world, as people struggle to secure their share of food, money, possessions, and status. They want to promote the well-being of their physical body, to enhance its charm and beauty, and to protect it from the threat of injury and death.

At the physical level consciousness is the binding force that keeps Nature intact and unites humanity with all living things. When we feel connected to the life-forms enveloped in the ecosystem, we are feeling the flow of vitality that binds Nature together.

It operates silently, without a voice, as it organizes the myriad of functions of the body. Yet even there, if we look at what is happening at the cellular level, it turns out that consciousness transcends all boundaries. Cells cooperate, communicate, exchange functions, perform acts of self-sacrifice, remain in balance, keep aware of their environment, adapt to change, and know that they survive by being part of a greater whole.

If we look at the physical worlds, it is obvious that our bodies are isolated from one another, which keeps us in separation by giving rise to the illusion that we must struggle and compete with every other isolated body. Yet it is through cooperation, physical security in social groups, and shared desire for food, shelter, sex, and physical comfort that brings us closer to wholeness.

At this level separation still seems to dominate over wholeness, which is why human beings continue to depredate the ecosystem without realizing that they are destroying part of their own life sustaining system. Separation occurs through imbalance, a disrupted ecosystem, pollution, and urban overcrowding.

Wholeness is achieved through vitality, kinship with other living things, balance in the ecosystem, and empathy. We all carry in our minds a large database of information that we consider fundamental. This database holds everything important that we believe about our world. We depend upon it to survive for even a short time.

Beliefs evolve over centuries, and therefore some

researchers look upon beliefs as being like virtual genes, that become fixed characteristics of the brain. These mental genes are often compared to a virus that spreads from person to person until the entire society gets infected. It would not be beneficial if our species become infected by everything. If we were actually open to all new ideas, we would not be able to maintain a coherent view of the world. In order to evolve, human beings had to make sure that they accepted only good ideas that promoted a coherent, reliable world-view – and rejected bad ideas, that moved the mind in the opposite direction.

The human brain processes only a fraction of the information available to it, and what we notice is not the same as what we know. Indeed, the entire world we inhabit of words and thoughts creates infinite brain changes in all of us every moment. In other words the processes and reflections may change brain cells.

This is exactly what was predicted by the new theory of the quantum mind. If mind comes before brain, then mind belongs to all of us? We can say my brain, but we cannot say my mind or my quantum field. There is growing evidence that in fact we do share the same mind field and we then participate together in a field, sometimes quite mysteriously.

Self-Awareness

Consciousness and self-awareness are, in the West, widely esteemed as the essence of being human, although the absence of self-awareness is considered a state of grace and perfection in the East. The origin of consciousness is imagined to be an unfathomable mystery, or the consequence of the insertion of an immortal soul into each human being, but not into animals, at the moment of conception.

Consciousness may not be so mysterious a trait, though, that supernatural intervention is needed to explain it. If its essence is a lucid awareness of the distinction between the inside of the organism and the outside, between it and everything else, then, most micro-organisms are to this degree conscious and aware – then the origin of consciousness on our planet dates back more than 3 million years.

There were vast numbers of microscopic creatures then, each with a rudimentary consciousness – perhaps only a micro-consciousness.

Consciousness and Social Interaction

If we look up consciousness in the dictionary, the definition is awareness of the world around us. Micro-organisms or bacteria are certainly aware – they interact, adapt, sense, respond, communicate, will orient themselves, work together to make structures, act creatively and respond specifically to the environment.

We cannot be aware of something without interacting with it somehow, anything we are aware of we are interacting with. For example, we cannot think of someone else without their image directly interacting with our thoughts.

How do our thoughts become reality? We do not think our thoughts but perceive them. Thoughts are sensations, like smells or visual stimuli. Our thoughts are the logical result of the relationships in our perspective. When we interact with something we are sharing a relationship with it. Consciousness is interaction.

Research shows that bacteria are not simple solitary organisms, or low level entities as earlier believed – they are highly social and evolved creatures. They consistently foil the medical community as they constantly develop strategies against the latest antibiotics. In the West, bacteria are one of the top three killers in hospitals today.

Recently published studies in the journal of BMC Genomics show that everyday pathogenic bacteria are not as smart – their IQ score is just about average, but **the intelligence of the Vortex bacteria (*Paenibacillus vortex*) is at the genius range, if compared to human IQ scores it is about 60 points higher than the average IQ of 100.**

Paenibacillus vortex is a species of pattern-forming bacteria, first discovered in the early 90's. It is a social microorganism that forms colonies with complex and dynamic architectures. Bacteria belonging to this genus have been detected in a variety of complex environments, such as soil, water, vegetable matter, forage and insect larvae, as well as clinical samples.

Social Intelligence Expressed by Microbes

All microbes, from bacteria to protozoa, make decisions throughout their life-span. They sense, interpret, and react to changing internal hemostatic states and local ambient environments, often staying with the same strategy or switching between alternative strategies of differential fitness to determine, for example, vegetative and reproductive cycles, phenotype, motility, and social-like cell to cell interactions. Successful strategies can increase a cell's viability and/or fecundity and may vary with inherited life-history traits, random or directed mutations, epigenetic modifications, and traditional forms of dual-process non-associative and associative learning and memory.

Strategy acquisition, modification, selection, and execution by microbes frequently require the coordinated workings of sensory transduction pathways, gene regulatory networks, membrane and intracellular transport systems, and motility and adhesion apparati.

The nano-scale physical dimensions of each of these components of a cell's computational machinery indicate microbial decision making operates within quantum mechanical limits even at biologically relevant temperatures and times (cf. Clark, 2010a, b)

Microbes may act alone or in concert with other microbes and organisms to overcome ecological barriers to survival and reproduction. In many respects, the cooperative and competitive social behaviors exhibited by microbes, such as assisted reproduction, altruistic suicide, reproductive cheating, quorum sensing, kinship recognition, induced defenses, and foraging (cf.Crespi, 2001), resemble rudimentary animal social intelligence.

Under natural growth conditions, bacteria live in complex hierarchical communities. To conduct complex cooperative behaviors, bacteria utilize sophisticated communication to the extent that their chemical language includes semantic and even pragmatic aspects. The complex colony forms emerge through the communication-based interplay between individual bacteria and the colony.

Individual cells assume newly co-generated traits and abilities that are not pre-stored in the genetic information of the cells, that is, not all the information required for efficient responses to all environmental condition is stored.

To solve newly encountered problems, they assess the situation via a collective sensing, recall stored information of past experiences, and then execute distributed information processing, **transforming the colony into a super-brain.**

Illuminating examples of swarming intelligence of live bacteria in which they solve optimization problems that are beyond what human beings can solve have been shown. Bacteria do not just react to change in their surroundings – **they anticipate and prepare for it.**

Quorum Sensing is a kin-recognition mechanism which bacteria use to implement density-dependent strategies. Each bacteria that utilizes Quorum Sensing produces a small molecule specific to its species called an auto-inducer and each bacterial cell both produces and is sensitive to this chemical. If many cells of one species are living in an area, there will be a large build-up of these auto-inducer molecules. The receptors of each cell are capable of measuring the presence in their environment, determining the density of their own species cells. When one bacterial species and its auto-inducers reach a certain concentration, the behavior of the bacterial population changes.

Ornstein, Robert (1986) developed the concept of the *multi-mind*. He identified many talents, such as activating, informing, smelling, feeling, healing, moving, locating and identifying, calculating, talking, knowing and governing. Could these apply to microbes?

Cannibalism among Microbes

Cannibalism among bacteria is a strange co-operative behavior elicited under stress. In response to stressors such as starvation, heat shock and harmful chemicals, the bacteria reduce their population with a chemical that kills sister cells in the colony. It works in much the same way that organisms reduce production of some of their cells when under starvation. But what is most interesting among bacteria is that they appear to develop a rudimentary form of social intelligence, reflected in a sophisticated and delicate chemical dialogue conducted to guarantee that only a fraction of the cells are killed.

In a current study, the researchers investigated what happens when two sibling colonies of a special strain of bacteria are grown side by side on a hard surface with limited nutrients. Surprisingly, the two colonies not only inhibited each other from growing into the territory between them, but induced the death of those cells close to the borders, researchers found.

Even more interesting to the scientists was the discovery that cell death stopped when they blocked the exchange of chemical messages between the two colonies. It looks as if a message from one colony initiates population reduction in the cell across the gap. Each colony simultaneously turns away from the course that will bring both into confrontation.

In only a year, bacteria can develop resistance to a new drug that may have taken years and a small fortune to develop, but drug developers have not yet utilized bacteria's cooperative behavior and social intelligence.

Bacterial social intelligence, conveyed through advanced chemical language, allows bacteria to turn their colonies into massive brains that process information, learning from past experiences to solve unfamiliar problems to better cope with new challenges.

If we want to survive the challenges posed by bacteria, we must first recognize that bacteria are not the simple, solitary creatures of limited capabilities they were long believed to be.

Researchers have revealed new details about how cannibalistic bacteria identify peers suitable for consumption. It is a step toward a broader effort to map all signaling molecules between organisms. These are the molecules that control biology.

Bacterial cannibalism occurs when a sub-population of a microbial colony eats another sub-population, harvesting the latter's nutrients to sustain growth in times when external food sources are limited. The phenomenon is well-known, but not well-characterized. For example, researchers have not known exactly how microbes identify, select and kill their genetically identical siblings.

The molecules that the bacteria uses to differentiate themselves are akin to those of a multi-cell organism, even though the microbes are genetically identical. Scientists also explored whether the molecules might be effective weapons against human pathogens.

Chapter 22
MICROBIAL COMMUNICATION

The Language of Microbes

Scientists are on the verge of decoding the special chemical language that bacteria use to talk to each other. Microbes release small molecules that enable millions of individuals in a population to coordinate their behavior. Disease causing bacteria use this language to decide when to infect a person or other host. Decoding the structure and function of compounds involved in this elaborate signaling process, known as quorum sensing , could lead to new medicines to block the signals and prevent infections.

For many years, bacteria cells were considered primarily as selfish individuals, but in recent years, it has become evident, that, far from operating in isolation, they coordinate collective behavior in response to environmental challenges, using sophisticated intercellular communication networks. Cell-to-cell communication between bacteria is mediated by small diffusible signal molecules that trigger changes in gene expression, in response to fluctuations in population density. This process, generally referred to as quorum sensing (QS) controls diverse phenotypes in numerous bacteria.

Recent advances have revealed that bacteria are not limited to communication within their own species but are capable of listening in and broadcasting to unrelated species to intercept messages and coerce co-habitants into behavioral modifications, either for the good of the population or for the benefit of one species over another.

Quorum Sensing

It is also evident that quorum sensing is not limited to the bacterial kingdom. The study of two-way intercellular signaling networks between bacteria and both unicellular and multicellular eukaryotes as well as between eukaryotes is just beginning to unveil a rich diversity of communication pathways.

In general quorum sensing systems facilitate the co-ordination of population behavior to enhance access to nutrients or specific environmental niches, collective defense against other competitor organisms or community escape where survival of the population is threatened. Although quorum sensing has primarily been studied in the context of single species, the expression of QS systems may be manipulated by the activities of other bacteria within complex microbial consortia which employ different quorum sensing signals.

These signal molecules also exhibit biological properties far beyond their role in co-ordinating gene expression in the producer organism. Consequently both quorum sensing systems and quorum sensing signal molecules have attracted considerable interest from the biotechnology, pharmaceutical and agricultural industries, particularly with respect to quorum sensing as a target for novel anti-bacteria.

Perhaps the most studied examples are in the expression of virulence factors. These are proteins that help bacteria to manipulate and evade its host. Expressing a virulence factor antagonizes the immune system, and makes it more probable that the bacterial population will come under attack, since host immune systems tend to evolve recognition of virulence factors. If the bacterial population expresses virulence factors when it is small, it is easily overwhelmed by the immune response. But if the bacteria population waits until it has sufficient numbers, it can overwhelm the host by switching en masse to the new pathogenic form, with its much greater effect.

Virus Invasion

Like a sleeper agent, the flu virus causes damage from within, turning organism's cells against itself. A single virus can hijack a healthy cell and transform it into a virus factory, making thousands of copies in a couple of hours. The cell then bursts, allowing the copies to infect other healthy cells and start the process anew. The body fights back by launching a self-sacrificing counter attack: molecules (cytokines) and T-cells designed to kill the hijacked cells before the virus does. These defenses come with a cost.

Most of the symptoms we attribute to a flu virus are actually the result of the body's defensive maneuvers. After an exposure to a cold or flu virus, the steps taken once infected, will determine whether the symptoms will be mild and temporary, or one will become moderately to seriously ill for days or weeks.

For example, fever occurs when cytokines tell the brain to raise the body's temperature, which helps the immune system to fight its enemies. **The clash between a viral infection and the body's immune defenses is one of nature's most dramatic conflicts,** an all-out battle with cells of the lungs, stomach or sinuses as the innocent casualties. Almost always, the immune system wins, successfully vanquishing the virus after three or four days of fever, coughs and aches.

Human or Microbe

The main reason why bacteria invade the bodies of human beings and other animals is because those bodies harbor environments where the bacteria can survive and multiply. Body fluids, such as plasma, are rich in sugars, vitamins, minerals and other chemicals which bacteria can use as nutrients.

Strictly by the numbers, the vast majority estimated by many scientists is that 90 percent of the cells we think of as our body are actually bacteria, not human cells. In fact most of all life on this planet is probably composed of bacteria. But as researchers describe in a series of recent findings, bacteria communicate in sophisticated ways, take concerted action, influence human physiology, alter human thinking and work together to bio-engineer the environment.

These finding may foreshadow new medical procedures that encourage bacterial participation in human health. They clearly set out a new understanding of the way in which life has developed on Earth to date, and of the power microbes have to regulate and manipulate both the global environment and the internal environment of the human beings they inhabit and influence so profoundly.

In short, the microbes that co-inhabit our bodies show considerable self-restraint by moderating the virulence of disease, especially in well-established relationships with hosts. Systemic pathogens such as staphylococcus or streptococcus that long ago invaded and live within our bodies, rarely secrete extreme toxins. Some of these one-time invaders have become permanently established in our cells, even crossed the boundary line and entered our own genome.

The implications of our new understanding are that we need more research, not only on how bacteria are virulent, but how they withhold their virulence and moderate their attacks. We need to investigate how our micro-biome flora, the ones we live with all the time, do not cause disease and instead protect us against their competitors.

View of Self

The true value of a human being
can be found in the degree to which he has
attained liberation from the self.

- Albert Einstein

For years our traditional view of self was restricted to our own bodies, composed of eukaryote cells encoded by our genome. However, in the era of gnomic technologies and systems biology, this view now extends beyond the traditional limitations of our own core being to include resident microbial communities, who outnumber our own cells and in exchange for food and shelter, these symbionts provide us, the host, with metabolic functions far beyond the scope of our own physiological capabilities. In this respect the human body can be considered a super-organism; a communal group of human and microbial cells all working for the benefit of the collective, a view which most certainly attains liberation of self.

The Missing Link

Microbes were the first form of life on this planet and have survived by adapting to ever changing environments, from an anaerobic life form to a photo-synthesizing and energy producing micro-organism to intelligent, decision making, life sustaining species in charge of many primary functions in Earth's biochemical and biological balances.

These transformations are in part due to the cooperative qualities among individual bacteria, being at the core of their social behavior and while colonies feature a division of labor and provide critical services to their community, they also congregate in immense numbers to fend off enemies, meet the challenges of nature to reproduce, obtain food and maintain their critical environments.

Microbes however, have survived and evolved, they sense, interpret and react to any changes in the environment, to challenges by invaders and may anticipate adverse conditions by executing intricate transmutation of their species. Microbes continuously sense and exchange messages within and among bacterial colonies and impart meaning that changes according to the environment.

Meaningful communication of this type fosters intentional behavior by bacteria, such as emitting and receiving pheromones in a courtship process before mating. They exhibit fundamental forms of social intelligence that have traditionally been viewed as solely human.

Sensing their environment, directly affects the bacterial behavior and ability to adapt. They are sensitive, communicative and decisive organisms and bacterial responses are more flexible, complex and adaptable than generally believed. In terms of re-defining cognition, behavior at the microbial level is precisely what must be understood in order to comprehend how more complex and specialized forms evolved and now function, and how cognition is part of basic biological functions.

Many single-celled organisms exhibit intelligence of a kind not seen in other species of the animal kingdom, they neither have nervous systems nor brains and are therefore not capable of conscious thoughts, but they harbor an internal system that can be **equated to a biological computer.**

From previous studies we know that the signaling molecules use to coordinate quorum sensing come in two categories:

One type of signaling molecule participates in chemical communication systems that are unique to each bacterial species.

The second type of signaling molecules occurs in all bacteria and fosters communication across species and may activate genes that contribute to virulence as well as swarming, bio-film formation, and other social behaviors.

Yet another type of communication may be through the quantum consciousness providing critical mass as needed.

At times, entire colonies may invade an unsuspecting host before transforming into vicious attackers. For instance, virulent species may lie dormant in a host until they reach critical density for releasing poisonous substances and resisting immune responses. Researchers' today suspect that quorum sensing may transform various bacteria into infectious killers.

If we humans are to survive as a species, it is imperative that we come into a new and very different relationship with the most prolific, most ancient, and possibly the most mysterious species on this planet. Re-generating our individual bio-terrain means forming alliances, not antagonisms, with the microbial realms. Instead of antibiotic ideology, we must espouse a pro-biotic (pro-life) philosophy of medicine.

Human Entanglement with Microbes.

The number of bacteria living within the human body of the average human adult are estimated to outnumber human cells 10 to 1 and are found mostly in the digestive system – the gut flora. This gut flora has a dynamic and continuous effect on the host's systemic immune system.

In order to understand how changes in normal bacterial populations affect us or are affected by disease we must consider lifestyle, nutrition, personal hygiene, exposure to stress, pollution and the environment. Just as the human mind allows a person to develop a concept of intellectual self, the immune system provides a concept of biological self.

Exposure to bacteria and/or viruses and our reaction to these invaders or their transmutations will largely depend on the health of our microbial environment and our mental, emotional state. **Microbes in our gut are connected to our brain, our brain embraces the mind and the mind interacts with quantum consciousness to communicate with microbes.**

Whether we study the research of Dr. Hamer, Hulda Clarke, Royal Rife, Wilhelm Reich and others, the conclusion that bacteria and viruses play a significant part in our physical well-being cannot be denied and the interaction between our physical and mental/emotional states is well documented.

How then can we explain disease caused by mental/emotional origin? For example: As in Dr. Hamer's case, the brutal shock inflicted by the death of his son Dirk and the consequent DHS in the brain would have most likely created a subconscious connection to his thought patterns. Being pre-occupied with these thoughts would have re-enforced the energetic footprint and given mass to the impact it had on his psyche. This critical mass was provided by bacteria, creating a physical connection via prostate cancer. In the event that the conditions were favorable to a resolution then a tipping point or healing crisis could have been reached and the conflict may have been resolved, by experiencing a quantum leap (change to a benign state, or spontaneous healing) or the conflict is maintained by engaging in continuous entanglement.

Microbes are the messengers of the mind responding to quantum consciousness subject to the law of attraction.

In everyday life microbes may be responsible for many major advances in technology, biology, chemistry, the arts, and human evolution by providing critical mass to inspiring a-ha moments (epiphany) as the creative force connecting the human mind to integrate with quantum consciousness, **being one with all knowledge, wisdom and intelligence of creation.**

Our life begins with microbes
Our well-being depends on microbes
Our life ends with microbes

- Helga Zelinski PhD

Citations

Abel, David Dr. 2010. *Constraint vs. Control. Controls are frequently confused with Constraints. The two are not synonymous. The concept they describe have little in common...* Updated 1 December 2010. Viewed 26 January 2011. <http://www.scitopics.com/constraint-vs_controls.html>

Agarwal, Ankit Scientist. 2009. *Silver in Tiny Doses Kill Bacteria,* Helps Wounds Heal...D.C. scientist Ankit Agarwal revealed that he s come up with an approach to....11 September 2009. Viewed 24 December 2010. <http://www. naturalnews.com/027010_silver_bacteria_medicine.html>

Agricultural field system *generally refers to innovative elements of prehistoric and historic agriculture is a variably complicated progress...* Updated 26 August 2010. Viewed 16 January 2011. <http://microbewiki.Kenyon.edu/ index.php.Agricultural_field>

Amazing Gut-Brain connection. 2010. *The connection between the brain and the gut is so strong that a new field of ...The correct microbiological measure for the live bacteria is ...*9 November 2010. Viewed 31 January 2011. <http://www.articlebase.com/.../the-amazing-gut-brain-connection3632827.htm>

Amazing patterns. 1995. *Amazing patterns of bacterial communication. The beautifully intricate patterns that colonies of bacteria form when....*4 March 1995. Viewed 6 February 2011 from <Bacterial Chatter, Science News, Mar. 4, 1995, p. 136-142>

Amid the Murk. *Gut Flora, vitamin D receptor emerges as a key player.* 2010. University of Rochester Medical Center, 8 July 2010. Science Daily. Viewed 11 December 2010. <http: / /www. sciencedaily.com / releases / 2010 / 07 / 1007071 41558.htm>

Anaerobic bacteria culture is a method used to grow anaerobes from a clinical specimen. *Obligate anaerobes are bacteria that can live...*October 13, 2010. Viewed 4 February 2011 <http: / /www. surgeryencycolpedia.com. A-Ce>

Ancient Health. Plate-of-raw-food New Year's resolution ....2008. Think about contaminated meat with Salmonella or E-Coli bacteria. ... It's significant that it wasn't until the quarantine practices laid out in the Bible were ... 28 December 2008. Viewed 16 December 2010 <http: / / random1881.wordpress.com / 2008 / 12 / 28 / ancient-health / >

Antibacterial. *An antibacterial is a compound or substance that kills or slows down the growth of bacteria. The term is often used synonymously with the term antibiotic....*Updated 2 February 2011. Viewed 10 February 2011. <http: / / en.wikipedia.org / wiki / Antibacterial>

Antioxidants. *The process of oxidation in the human body damages cell membranes and other structures including cellular proteins, lipids.....*Viewed 19 February 2011. <http: / / www.betterhelth.vic. gov.au / bhcv2 / bhcarticles... / Antioxidants>

Archibald, J. How do microbes rule the world? *...More than 90 percent of the cells in the human body are not....*Viewed 21 January 2011. <http: / / www.cifarnbq.ca / questions / how-do-microbes-rulethe-world.commtechlab.mso.edu / news / ncancient.html>

Are Bacteria Intelligent? *Are you ready for this? Some scientists are now thinking that it might be possible to find intelligent bacteria someday....*Viewed 21 December 2010. <http: / / aboutfacts.net / science178.html>

Arsenic-Eating Bug. 2010. *Scientist dabbling in the toxic waters of a California lake have found a new species of bacteria that thrives on poisonous arsenic* a...2 December 2010. Viewed 3 December 2010. <http://www.aolnews.com/.../scientists-find-arsenic-eating-bug-that-holds-cluesto-extraterr/>

Artificial Intelligence. 2009....*Artificial Intelligence is a multi-disciplinary field, which involves at least psychology, cognitive science, computer science and...* 18 December 2009. Viewed 4 February 2011. <http://www.anapsid.org/her-intelligence.html>

Atkinson, Steve; Williams, Paul. 2009. *Quorum sensing and social networking in the microbial world. For many years bacterial cells were considered primarily selfish individuals.* Viewed 29 December 2010. <http://rsif.royalsocietypublishing.org/content/6/40/959.abstract>

Ayurveda. *Considered by many scholars to be the oldest healing science, Ayurveda is a holistic approach to health that is designed to help...*Viewed 8 March 2011. <http://www.umm.edu... Complementary Medicine>

Ayurveda, *the complete knowledge for long life or Ayurvedic medicine in a system of traditional...*Viewed 8 March 2011. <http://en.wikipedia. org/wiki/Ayurveda>

Bacteria. *How do Bacteria Divide and Multiply? Bacteria are single celled organisms...*Viewed 22 January 2010. <http://ict-science-to-society.org/Pathogenomics/bacteria.htm>

Bacteria and Archaea – Who's there? Microbial Lifestyles... *Humans and other animals therefore have two distinct single-celled ... Microbiologists estimate that a mere 1 percent of the microbes that are visible ... The field of comparative genomics is filled with efforts to compare gene* ...Viewed 15 February 2011.<http://science.jrank.org/pages/48315/Bacteria-Archaea.html>

Bacteria are more capable of Complex Decision-Making Then Thought. *2010. It's not thinking in the ways humans, dogs or even birds think, but new findings* from researchers at the University of Tennessee, Knoxville, show... 19 January 2010. Viewed 3 February 2011. <http://www.sciencedaily.com/releases/2010/01/100114143310.htm>

Bacteria - *Characteristics of Bacteria*.....Viewed 15 December 2010. <http://science.jrank.org/pages/714/ Bacteria.html> Bacteria - Characteristics of Bacteria

Bacteria: enemies or friends? *We try our best to destroy bacteria and live without them - in the mistaken belief that doing so will rid us of all infectious*...Viewed 8 March 2011. <http://www.natural-cancer-cures.com/ bacteria.html>

Bacteria Far More Intelligent Than We Realize? 2010. *Strictly by the numbers, the vast majority estimated by many scientists at 90 % - of the cells in what you think*. 11 November 2010. Viewed 28 December 2010. <http://www.disinfo.com/2010/11/are-bacteria-far-more-intelligent-than-werealize/>

Bacteria in the Gut May Influence Brain Development. 2011. *A team of scientists from around the globe have found that gut bacteria may influence mammalian brain development*..... 1 February 2011. Viewed 21 February 2011. <http://www.sciencedaily.com/releases/2011/01/110201083928.htm?>

Barefoot Doctor's Manual - 1970 - *A Guide to Traditional Chinese and Modern Medicine. Coles Medical Books. This medical manual is an expression of Modern China s intense effort to remember its past, to use what has been shown to be effective......Institute of Traditional Chinese Medicine of Hunan Province in China (because they worked in the paddy fields like any other commune member, barefooted and with trouser legs rolled up, they were given the name barefoot doctors* Peking Review 1977). Preface.

Bassler, Bonnie. Microbiologist. 2009. *Brain, Mind, Consciousness and Learning. How bacteria communicate.* 9 April 2009. Viewed 8 December 2010. <http://brainandlearning.com/2009/04/howbacteria-communicate.html>

Becchetti, Sonia Kessler, Michael. 1999. *Biological Terrain and Assessment: an Invaluable Tool...measures blood, saliva and urine for electron levels, ph balance and minerals in these fluids. The best comparison...* 1 November 1999. Viewed 13 February 2011. <http://www.dynamicchiropractic.com/mpacms/dc/article.php?id=36339>

Belrad, Bryan. *Does intestinal bacteria cause chocolate cravings?* On Friday, October 19th, a new study was published in the Journal of Proteome. Viewed 4 December 2010. <http://helium.com/items/658831-does-intestinal-bacteria-cause-chocolate-cravings>

Bendston, William Ph.D., Fraser, Sylvia. 2010. *Chasing the Cure.* Key Porter Books Ltd. Toronto, Ontario

BENEFICIAL AND EFFECTIVE MICROORAGANISMS FOR SUSTAINABLE AGRICULTURE AND ENVIRONMENT.... *UTILIZATION OF BENEFICIAL MICROORGANISMS IN AGRICULTURE* Contents viewed 28 January 2011. <http://www.agriton.nl/higa.html>

Beneficial microbes in agriculture / *Shop for natural probiotic agricultural products to increase soil fertility and water holding capacity using beneficial...*Viewed 31 January 2011. <http://www.scdprobiotics.com/ Agriculture's/311.htm>

Benton, D. 2006. *Impact of consuming milk drink containing a probiotic on mood and cognition.....*6 December 2006. Viewed 14 December 2010. <http://www.ncib.nkm.nih.gov/pubmed/17115194>
Billion-year-old land dweller. *Fossils Hint Land Life Began in Far Earlier Era,* Viewed 2 January 2011. <http://mucrobezoo.com>

Biological Warfare (previously called germ warfare) 2010.... *Is the use of disease causing microorganisms as military weapons* ...22 January 2002. Viewed 1 February 2011. <http://www.science-clarified.com - Bi-Ca>

Biological warfare, *also known as germ warfare, is the deliberate use* ...Zelicoff, Alan and Bellomo, Michael (2005), Microbe: Are we Ready for the...Viewed 1 February 2011. <http://en.wikipedia.org/wiki/Biological_warfare>

Blood - Holograph of Your Consciousness. *A Reflection of Who You Are. Peering into the microscope and looking at live blood, you can see a reflection*....Viewed 10 February 2011. <http://biomedx.com/microscopes/rrintro/ rr5.html>

Body Ecology. *If you ever had diarrhea, then you know how uncomfortable the frequent, loose stools can be. Even worse, they are often accompanied by...* Viewed 27 February 2011 <http://bodyecolgy.com/articles/probiotic_ diarrhea.treatment.php>

Borenstein, Seth. 2007. Scientists *Explain Chocolate Cravings*. By Seth Borenstein. The Associated press Friday, 12 October 2007; 4:34 AM. Washington If that craving for... Viewed 26 October 2010. <http://www.washingtonpost. com-Health-Wires>

Borlongan, J. *How do bacteria multiply? Binary fission is an asexual reproduction method*....Viewed 22 January 2011. <http://www.ehow.com/how-does-4565349_bacteria-multiply.html>

Bower, Bruce. One-Celled Socialites. *Bacteria mix and mingle with microscopic fervor. Welcome to a vibrant social scene that has operated largely in secret....* Viewed 20 November 2010. <http://www.phschool.com/ science/ science_ news/articles/one_celled_socialites.html>

Bowsher, Clive G. 2010. Information processing by biochemical networks: a dynamic approach. *Understanding how information* *is encoded and transferred...*4 August 2010. Viewed 12 December 2010. <http://rsif. royalsocietypublishing.org/content/early/2010/.../rsif.2010.0287.abstract>

Boyle, Rebecca. 2010 *Quantum Entanglement May Hold Human DNA Together. DNA Double Helix. A new study suggests quantum entanglement might be responsible for holding DNA together*. 28 June 2010. Viewed 25 December 2010. <http://popsci.com/.../quantum-entanglement-may-hold-dnatogether-new-study-says>

Brain-Gut Connection. *Why do we get butterflies before a performance? The answer may surprise you...*Viewed 31 January 2011. <http://altmedangel.com/gutbrain.htm>

Brain, Mind, 2009. Consciousness and Learning: *How Bacteria Communicate...In quorum sensing, bacteria assess their population density by detecting the concentration of a ...* 9 April 2009. Viewed 12 December 2010. <http://brainandlearning.blogspot.com/.../how-bacteria-communicate.html>

Bransfield, Robert, CMD. Microbes and Mental Illness. *Microbes are the greatest predator of man. As medical technology improves, there is increasing ...* Viewed 4 January 2011. <http://www.mentalhealthillness.com /.../MicrobesAnd MentalIllness.htm>

Brief History of Dr. Royal Rife. *Dr. Royal Rife was a brilliant scientist and a hero in our time. He developed many a scientific breakthrough in both the field of microscopy as...*Viewed 2 January 2011. <http://www.quantumbalancing .com/rife-historyif.htm>

Brief History of Microbiology. *Microbiology has had a long, rich history, initially centered on the causes of infectious diseases.....*Viewed 4 December 2010. <http://www.cliffsnotes.com/study_guide/topicArticled-8524,articled-8406.htm>l

Bugs and Brains: Gut Bacteria Affects Multiple Sclerosis. 2010. *Biologists at the California Institute of Technology have demonstrated a connection between*...20 July 2010. Viewed 28 December 2010. <http://sciencedaily.com/releases2010/07/100719162643.htm>

Bugs Inside; *What Happens When the Microbes That Keep Us Healthy***.** 2009... And, when he died last year, we had to beat...16 December 2009. Viewed 6 February 2011. <http://www.scientificamerican.com/article.cfm?id =human...change>

Cannibalism. 2009. *Cannibalism among bacteria, explains prof. Ben Jacob, is a strange cooperative behavior elicited under stress. In response to stressors such*...2 February 2009. Viewed 22 November 2010. <http://esciencenews.com/articles/2009/02/.../can.cannibalism.fight.infections>

Cantwell, Alan, M.D. 2007. All Human Blood is Infected with Bacteria. Bacteria are everywhere. Our mouths, nose, ears all harbor germs....Viewed 17 February 2011. <http://www.rense.com>

Case, Christine L., Ed. D. Microbiology of Chocolate. *Cacao seeds must be fermented, dried, and roasted to produce the chocolate flavor.* Viewed 8 March 2011. <http://www.smccd.edu/accounts/case/chocolate.html>

Causes and Consequences for Human Substance...2010. *Social forces that motivate emigration and immigration can vary from... While their host develops immunity to a particular microbe over*...30 March 2010. Viewed 27 January 2011. <http://fubini.swarthmore.edu/-ENVS2/mperch1/Mperchessay3.html>

Cavalier-Smith, Thomas; Brasier, Martin; Embley, T. Martin 2006. Introduction: *How and when did microbes change the world?* Phil. Trans. R. Soc. B 361 (1470): 845-50.Doi: 10, 1098/rstb.2006.1847. PMID 16754602

Chisholm, Hugh, Ed (1911) Encyclopedia Britannica (Eleventh

ed.) Cambridge University press. *Quarantine is compulsory isolation, typically to contain*...Modified 18 January 2011. Viewed 1 February 2011. < http://en.wikipedia.org/wiki/>

Chocolate cravings may be a real gut feeling. 2007 - Health - Diet and... *if that craving for chocolate sometimes feels like it is coming from deep in your gut, that s because maybe it is.* A small study links the...12 October 2007. Viewed 4 December 2010. <http://www.msnb.msn.com/id21257175/ns/health-diet_and_nutrition/>

Chopra, Deepak M.D. 1990. *Quantum Healing. Exploring the Frontiers of Mind/Body Medicine.* Bantam Books. New York. United States of America pg. 16-106

Chopra, Deepak M.D. 1994. *Ageless Body, Ageless Mind.* Three Rivers Press, New York

Chopra, Deepak. 2006. *Life after death: The burden of proof.* 1st Ed. Three Rivers press, New York, USA [pgs. 145-152, 219-228

Clark, D. 2009. Microbes. An Invisible Universe. *The ecology of microbes to one another and their surroundings is extraordinary with respect to the diversity of chemical and physical*.....22 August 2009. Viewed 27 December 2010. <http://ensobottles.com/blog/2009/08/microbes_au_invisible_universe/>

Clark, Hulda Regehr. PhD. ND. 1993. *The Cure for all Cancers. Including Over 100 Case Histories of Persons Cured.* New Century Press. United States. Reprinted 2006. Pages 10-11

Clark, Hulda Regehr. PhD. ND. 1995. *The Cure for All Diseases. With many Case Histories of diabetes, high blood pressure, seizures, chronic fatigue syndrome, migraines, Alzheimer's, Parkinson's, multiple sclerosis, and other showing that all of these can be simply investigated and cured.* New Century Press. United States. Reprinted 2005.

Clark, Hulda Regehr, PhD, ND. 1999. *The cure For Advanced Cancers. Including hard evidence of tumor shrinkage and disappearance.* New Century Press, United States. Reprinted 2005

Clark, Hulda Regehr. PhD. ND. 2004. *The Prevention of All Cancers. With Case Study for Syncrometer Testers.* New Century Press, United States.

Classification of Living Things. 2008. *With such diversity of life on Earth, how does one go about making*....Viewed 5 January 2011. <http://www.windows2universe.org/earth/Life/classification_intro.html>

Colgan, Michael Dr. 1996. *The New Nutrition. Medicine for the Millennium.* Apple Publishing Company, Hong Kong Comparing Chimp, Human DNA. 2006.

Most of the big differences between human and chimpanzee DNA lie in the regions that do not code for genes, according....13 October 2006. Viewed 15 December 2010.<http://www.sciencedaile.com/releases/ 2006/10/131 04633.htm>

Connections between human health and microbiology go all the way back to the discovery of microbes. The amateur Dutch scientist Antonie van Leewenhoek is...Viewed 6 January 2011. <http://www.bscs.org/curriculumdevelopment /highschool/.../humans.html>

Consciousness and Being Human - essays - Humanity without Physical...*We are intuitively aware of the result because the consciousness that has figured it out has formed a relationship with us. These microbes pass from*...Viewed 10 February 2011. http://www.ecsys.org/ecsys_consciousness_and_reality.php

Constipation. *Includes a definition of constipation and information on how it develops, how it is diagnosed, and how it can be treated. Also provides details on*... Viewed 13 February 2011. <http://digestive.niddk. nih.gov/diseases/pubs/constipation/>

Constipation Solutions. *A lack of beneficial microbes often result in constipation, gas, and bloating. In addition, HSOs work from the inside of the intestines...* Viewed 2 February 2011. <http://www.optimalhealthnetwork. com/Constipation-s/235.htm>

Controlling the Spread. 2006. Centers for Disease Control and Prevention - *Controlling the Spread of Contagious Diseases.... These are common healthcare practices to Control the spread of a contagious disease by.....* Viewed 1 February 2011. <http://www.redcross.org.preparedness/cdc/_english /IsoQuar.asp>

Corrin, Nicholas, L.Ac *Understanding Pleomorphism and Regenerating the Bio-Terrain. Louis Pasteur is the father, and in some sense the godfather, of modern germ theory......* Viewed 28 December 2010. <http://eclectichealing.com/pleomorohism.html>

Crawford, Dorothy H. 2009. Deadly Companions: *How Microbes Shaped Our History. Combining tales of devastating history. Deadly Companions reveals how closely microbes....*Viewed 4 December 2010. <http://www.goodreads.com/book/show/6411801-deadly-companions>

Crill, D. *Why milk microbes? Breast milk contains a great source....* Viewed 21 January 2011. <http://http://www.nutrinexus.com/about-nutrinexus>

Crill, Dick Ph.D. *Did you know there have been places in the world where people have regularly lived to be more than...*Viewed 6 February 2011. <http://www.nutrinexus.com/about-nutrinexus/>

Czaaraan, Tamaas L.Rolf, F.Hoekstra, Ludo Pagie ...2002. *Chemical warfare between microbes promotes biodiversity...* 22 January 2002. Viewed 1 February 2011. <http://www.pnas.org/content/99/2/786.abstract>

Day, Pamela Noeau. 2009. *Poi, Fermentation and Traditional Poi Cultures. Taro is an ancient food and was the staff of life for the Hawaiian people.* For centuries....23 March 2009. Viewed 3 February 2011. <http://www.poico.com/artman/publish/article_72.php>

Defecation, 16th century drawing of a person defecating in squatting position outside... *until the next mass peristaltic movement of the transverse and descending colon. ... Excessive pressure placed upon the abdomen, inflammatory bowel disease,...* Viewed 22 January 2011. <http://en.wikipedia.org /wiki/Defecation>

Difference between aerobic vs. anaerobic bacteria. *There are two types of organisms and tiny single celled bacteria called aerobic and anaerobic bacteria in the human body.* Viewed 14 November 2010.<http://www.differencebetween.net/science/difference-between-aerobic-and-anaerobicbacteria>

DNA Patterns of Microbes. 2009. *The genomes or DNA of microbes contain defined DNA patterns called genome signature. Such signatures may be used to establish relationships* ... 25 June 2009. Viewed 15 February 2011. <http://www.sciencedaily.com/releases/2009/06/090625074625.htm>

Dormant Microbes Promote Diversity, Serve Environment. 2010. *The ability of microbes, tiny organisms that do big jobs in our environment, to go dormant*...25 March 2010. Viewed 16 February 2011. <http://sciencedaily.com /relaeases/100322111943.htm>

Doyle, RJ, Lee. NC. 1986. *Microbes have played a profound role warfare, religion, migration of populations, art, and in diplomacy, Boundaries of nations have changed as a result*...Viewed 1 February 2011. <http://www.ncbi. nlm.gov/pubmed/3518891>

Druin, Paul MD PhD - 2008. *Five Pillars of Health. Bio-Terrain.* IQBNM. Updated 2010. Viewed 20 November 2010. <www.iquim.org>

Early Life -1998...*gradual changes in the earliest cells gave rise to new* *life forms...Earth's air would still be without oxygen and animals which need...*Viewed 26 December 2010. <http://www.windows2universe.org/.../tour.php? ... =/earth/Life/early-life.html>

Ehrenberg, R. 2010. *Baby s first bacteria depend on the birth route.* Viewed 7 November 2010. <http://sciencenews.org/view/generic/id/60461>

Eshel Ben-Jacob. 2009. *Learning from Bacteria about Natural Information processing. To further assess the extent of bacteria's social intelligence we......as A, B -using the quantum mechanics (quantum computing) notations of...*15 October 1999. Viewed 7 November 2010. <http://on;inelibrary. wiley.com/doi/10.1111/1.1749-6632.2009.05022.x/full>

Etris, Samuel. 2009. Norwex Eco Friendly green Cleaning Products - *Why Silver Kills Germs & Heals Wounds.*15 July 2009. Viewed 11 December 2010. <http://www.merchantcircle.com-Top-MN-Saint Paul - 55199>

Dead man decomposing. 2003. *Like many big-bellied people in Tennessee, the dead man is dresses for comfort....* 17 April 2003. Viewed 16 February 2011. <http://dir.salon.com/story/mwt/feature/2003/04 17/roach.../index. html>

*Decomposition is the process by which organic material is broken down...*Viewed 16 February 2011. <http://en. wikipedia.org/wiki/Decomposition>

Digestive Disease: *Disease is an abnormal condition affecting the body of an organism. It is often construed to be a medical condition associated with specific symptoms and ...* Viewed 27 January 2011. <http://en.wikipedia.org/wiki/Disease>

Digestive System. Digestion Introduction. *Just a spoonful of sugar...* goes the song. Viewed 11 December 2010. <http://www.webmed.com/digestive-disorders/digestive-system>

Digestive System Information. 2011. *The mouth is the beginning of the digestive system, and, in fact, digestion starts.....*Viewed 4 February 2011. <http:/ /www.medicinenet.com...the digestive system index>

DNA Patterns of Microbes. 2009 *The genomes or DNA of microbes contain defined DNA patterns called genome signature. Such signatures may be used to establish relationships ...* June 25, 2009 Viewed 15 February 2011. <http:/ /www.sciencedaily.com/ releases/2009/06/090625074625.htm>

Do Probiotics Influence Mood and Mental health? 2010. It may seem strangely related, but the benefits of probiotics have been extended to cognitive and emotional health... 4 November 2010. Viewed 18 December 2010. <http:/ /health /?p=553>

Does the Mind Survive Death? Does the conscious mind survive death? 2009. If so does it ALL survive death, or just our "core" self? ...11 February 2009. Viewed 16 February 2011. <http:/ / paranormal-light.blogspot.com/2009/02/does-mind-survive-death.html>

Druin, Paul M.D., Ph.D., *Quantum Microscopy*. International Quantum University of Integrative Medicine.

Druin, Paul MD PhD. *Five Pillars of Health....assimilation, elimination, oxidation, Immunity, regeneration....*International Quantum University of Integrative Medicine

EcoHeart - Effective. 2010. EcoHearth - Effective Microorganisms: Using Bacteria and Yeast to...*An intimate understanding of the local ecology is necessary for successful agriculture and it may be important to extend this knowledge to...* 27 August 2010. Viewed 28 January 2011. <http:/ /ecohearth.com/.../1496-effective-microorganisms-using-bacteria-and-yeast-to-createsustainable-agriculture.html>

Edmonds, Molly. 2009. Discovery Health. *The Body after Death. Take a look at what happens to the body after death, from death chill...A few days after death, these bacteria and enzymes start the*...12 January 2009. Viewed 16 February 2011. <http://health.howstuffworks.com/diseaseconditions/death-dying>

Eisenstein, Mayer. MD. JD. MPH. Spring 2008. *How to reduce digestive issues without lowering stomach acid. American Academy of Anti-Aging Medicine.* Anti-Aging medical News. AAMN and A4M Buyers Guide produced by MPA Media, Huntington Beach, CA pg. 11

Emotion and memory. *Emotion can have a powerful impact on memory. Numerous studies have shown that the most vivid autobiographical memories tend to be of emotional*....last modified 2 January 2011. Viewed 12 January 2011. <http://en.wikipedia.org/wiki/Emotion_and_memory>

Fecal bacterio-therapy, *also known as fecal transfusion, fecal transplant, or human probiotic infusion*...last modified 31 January 2011, Viewed 6 February 2011. <http:Wikipedia.org/wiki/Fecal_bacteriotherapy>

Fermented Rice Bran. *If beneficial microorganisms are not the result of the food we eat, then we must find food*.....Viewed 19 February 2011. <http://www.aliveandbewell.com/fermented_rice_bran>

Flu Virus. 2009. *A complete overview of the flue. Highlights all the most important aspects of the flu virus and what you need to know.* 25 June 2009. Viewed 5 March 2011. <http://coldflue.about.com/od/flu/p/fluprofile.htm>

Fossils Hint Land Life Began...1994. *The discovery of fossilized filaments from bacteria or blue-green algae indicates*...The New York Times, 22 February 1994, p. B7

Frank, Guenther. THE FASCINATION OF KOMBUCHA. Nowadays we experience a return to healing measures that are close to nature.....Viewed 21 November 2010. <http://www.acupuncture.com/herbs/kombucha.htm>

Frank, Jeff - QUANTUM PHYSICS and HORTICULTURE. *Thousands of years ago it was known that the Earth was BIG, however men of learning like the Greek ...* Viewed 16 December 2010. <http://www.thenaturelyceum.org /QUANTUM.html>

Gardiner, SM. 200. *Haemodynamic effects of bacterial quorum sensing signal, N-(3-oxododecanoy) L-homoserine lactone, in conscious, normal and endotoxaemic rats. ...* Viewed 10 December 2010. <http://www.ncbi.nlm.nih.gov/pbmed/11487515>

Genius of bacteria. 2011. IQ scores are used to assess the intelligence of human beings. Now Tel Aviv University has developed a social-IQ score for bacteria - and it may lead to new.....24 January 2011. Viewed 25 January 2011. <http://www.sciencedaily.com/releases/2011/01/11012411138.htm>

Gerson, Charlotte; Walker, Morton. 2001/2006. The Gerson Therapy - *The Proven Nutritional program For Cancer and Other Illnesses.* Kensington Publishing Corp. New York. NY. Pages 258-259

Gerson, Max. M.D. 2002. *A Cancer Therapy. Results of Fifty cases and The Cure of Advanced Cancer by Diet Therapy.* Gerson Institute, United States. Sixth Edition. Pg. 49, pg. 130-133

Gill, Zann. 2010. *If Microbes begat Mind. Collaborative Mind, collaborative intelligence... this was not a simple, linear sequence - evolution begat life, we begat intelligence.* Life.... 2008-2010. Viewed 4 December 2010. <http://zanngill.com/1mbm.html>

Gillen, A.L. 2008. *Microbes and the Days of Creation.* Viewed 17 December 2010. <http://www.Answers.genesis.org/articles/arj/vl/nl/microbes-days-of-creation>

Gladwell, Malcolm - 2002. *The Tipping Point*. Little Brown and Company, Hachette Book Groug, New York

Gladwell, Malcolm - 2008. *Outlier.* Little Brown and Company. Hachette Book Group. New York, NY

Glausiusz, J. 2007. *Your Body Is a Planet. 90% of the cells within us are not ours but*...19 June 2007. Viewed 11 September 2010. <http://discovermagazine.com/2007/jun/your-body-is-a-planet>

Going Mental - Big Think. 2010. *Einstein said: The true sign of intelligence is not knowledge but imagination. Socrates said: I know that I am intelligent*... 2 September 2010. Viewed 4 February 2011. <http://bigthink.com/ideas/23098>

Goldman, Armond S. and Prabhakar, Bellur S. *The human immune system is extremely complex. It has evolved over hundreds of millions of years to respond to invasion by the pathogenic* ... Viewed 19 February 2011. <http://crohn.ic/archive/PRIMER/immunesys.htm>

Goswami, Amit Ph.D. 2004. *The Quantum Doctor*. Hampton Road Publishing Company Inc. Charlottesville. VA

Gross, L. 2007. Microbes colonize a Baby's Gut with Distinction. *Where do the microbes come from? What effects*... Viewed 7 November 2010. <http://www.ncbi.nlm.nih.gov...PLoL Biol v.5(7); Jul 2007>

Gut-Brain Connection & Autism, ADD, Allergies, and Other ...2010. *Bad bacteria in the gut can emit toxins, which affect brain function... So that's my Gut/Brain connection overview, along with the Cara* ... 11 November 2010. Viewed 31 January 2011. <http://healthhomehappy.com/.../the-gutbrain-connection-autism-add-allergies-and-other-diseases.html>

Gut Brain Connection - Digestive Problems and Mental Health.

There is definite gut brain connection and problems associated with the ... Our digestive system contains billions of bacteria (or gut flora) that have ... Viewed 31 January 2011. <http://www.improve-mental-health.com/gut-brain.html>

Gut-Brain Connection. *Leaky Gut Syndrome or Hyper permeability ... allowing bacteria and bacterial fragments across the small intestinal lining.* Viewed 16 January 2011. ..<http://www.developmentaldelay.net/page.cfm/349>

Gut/Brain Connection. *Re-colonisation of the gut with beneficial bacteria is the aim following the removal of pathogenic bacteria.* ...Viewed 31 January 2011. <http://bedrockcommunity.org/idl5.html>

Gut Brain Connection. What is the gut brain connection? *Does it relate to migraine, or other headaches? ... of dangerous bacteria that are kept in check by the good guys.* ...Viewed 31 January 2011. <http://relieve-migraine-headache.com/the-gut-brain-connection.html>

Gut Bacteria Could be Key Indicator of Colon Cancer Risk. 2010. *The human body contains more bacteria than it does cells. These bacterial communities*.....24 June 2010. Viewed 11 December 2010. <http://sciencedaily.com/releases/2010/06/100622091738.htm>

Hamer's Biological Law #4:2010. *The Role of Microbes in Healing - You... Hamer's fourth biological law describes how microbes help heal...* Dr. Pedro Zayas M.D., F.A.A.F.P. ON Dr.Ryke Hamer's German New Medicine. 11 January 2010. Viewed 14 January 2011. <http://www.weinholds.org/.../hamers-biological-law-4-the-role- of-microbes-in-healing.html>

Hamer, Ryke, Geerd, Dr. Med - Hamer held a licence to practise medicine from 1963 until 1986 when it was....*Hamer maintains that these microbes, rather than being antagonistic to the*...Viewed 6 January 2011. <http://www.en.wikipedia.org/wiki/Ryke_Geerd_Hamer>

Hamer, Ryke Geerd, Dr. Med. Mag. theol. 2007.New German Medicine. *The five biological natural laws of German New Medicine.* Amici di Dirk, Ediciones de la Nueva Medicina S.L

Hamer, Ryke, Geerd Dr. Med. 2000. Summary of the New Medicine. *Presentation to comply with the qualification as lecturer of 1981, at the University of Tuebingen.* Amici Di Dirk - Ediciones de la Nueva Medicina S.L.; E-Fuengirola, Spain

Hamer, Ryke Geerd Dr. Med. – *The New Medicine, Mind Motivations.* Viewed 6 October 2013. <http://mindmotivations.com/resources/articles/new-medicine>

Handelsman, Jo. 2004. Metagenomics: *Application of Genomics to Uncultured Microorganisms. Metagenomics (also referred to as environmental and community genomics) is the genomic analysis.....* December 2004. Viewed 21 December 2010. http://www.ncbi.nlm.nih.gov/pmc/articles/PMC539003/>

Hanson, LA. 1998. *Breastfeeding provides passive and likely long-lasting active immunity. Allergy Asthma Immunol.* 1998 Dec; 81(6):523-33; quiz 533-4, 537. Viewed 21 January 2011. <http://www.ncbi.nlm.nih.gov/pubmed/ 9892025>

Harvest, fermentation, Drying and Transport of the Cacao Bean. *From the Tree to the Factory. A neatly packaged chocolate bar is the end result.* Viewed 8 March 2011. <http://www.allchocolate.com/ ...chocolate_is...tree_to_factory.aspx>

Heal Your Wounds With Sugar. *Everybody loves sugar. And here is another reason to keep it in your cupboard.* Sugar can heal... Viewed 11 December 2010. <http://www.naturalremediesblog.net/healyour-wounds>

Healing Cancer *Naturally: Dr. Ryke Geerd Hamer's German New Medicine integrates emotional cancer triggers and the inherent self-healing capacity of the body.* Viewed 22 January 2011. <http://www.healingcancernaturally. com/hamer.html>

History of Agriculture *Using minerals to rejuvenate soil is not a new idea....The microbes allow the nutrients to be biologically available...* Modified 25 January 2011. Viewed 29 January 2011 <http://en.wikipedia.org /wiki/History.of.agricultur >

Health; It all begins with pH 2010 *How You Rot & Rust....He observed microbe like particles in the blood which he called.....some level of microbial consciousness are actually behind aggressive,...*12 August 2010. Viewed 10 February 2011. <http://Ajp619.wordpress.com/2010/.../how-you-rot-rust-it-all-begins-with-ph/>

Health Benefits of the Natural Squatting Position. *Squatting prevents constipation in four ways: Gravity does most of the work. The weight of the torso presses against the thighs and naturally compresses the...* Viewed 19 January 2011. <http://www.naturesplatform.com./health_benefits.html>

Hill, David and Artis, David. 2009. *Maintaining Diplomatic relations between mammals and Beneficial Microbial Communities....* 24 November 2009. Viewed 23 December 2010. <http://stke.sciencemag.org/sgi/content/abstract/2 /98/pe77>

Hoffer, Abram M.D.; Walker, Morton D.P.M; 1996. *Putting it all together: The New Orthomolecular Nutrition.* Keats Publishing Inc. New Canaan, Connecticut

Honey heals your wounds. 2006. *Honey is more effective in treating difficult wounds than antibiotics, says....* 15 October 2006. Viewed 11 December 2010. <http://www.newmediaexplorer.org/sep/2006/10/15/honey-heals-your-wounds.htm-with-sugar-or honey/>

Horowitz, Leonard G. Dr.; Barber Joseph E. 1999. Healing Codes for the Biological Apocalypse. Tetrahedon Publishing Group. Sandpoint, Idaho

How Chocolate is made. *After picking, gatherers follow the harvesters who have removed the ripe pods from the trees. The pods are collected...*Viewed 8 March 2011. <http://www.waynesthisandthat.com. howtomakechocolate2.htm>

How honey heals wounds. *The problem with wounds. In normal situations, like a cut finger or a grazed knee, the wound will heal...* Viewed 11 December 2010. <http://www.biotechlear,org.nz/.../honey...heal /how_honey_heals_wounds>

*Human flora is the assemblage of microorganisms, benign and otherwise, that reside on the surface and in the deep layers of skin, in the saliva and oral...*Modified 28 September 2010. Viewed 22 October 2010.<http://en.wikipedia.org/wiki/Human_flora>

Human Lineage. 1999. *In a stunning announcement, a team of researchers revealed.* Viewed 12 December 2010. <http://biology.about.com/library /weekly/aa042999.ht>

Humans carry in their genome 2010. ..*The relics of an animal virus that infected their forerunners at least 40 million years ago, according to* ... 6 January 2010. Viewed 15 February 2011. <http://www.physorg.com/news182006019. html>

Human vs. Animal DNA. *How alike are we? Difference between Human and Animal DNA. At the center of every living creature on the planet is a blueprint- a map...*Viewed 28 December 2010 <http://recomparison.com/ comparisons/100668/human-vs-animal-dna-how-alike-are-we/>

Hustvedt. How does consciousness fit into this? ... *are there in a group------something called quorum-sensing-----and decide to attack or not based on the collective* ... Viewed 12 December 2010 <http://discover.coverleaf.com/ discovermagazine/201012?pg=78>

If You Have Certain Food Cravings. 2008.... *It May Be Due to the Bacteria... Sugar cravings ...We have all got them, but new research shows certain strains of bacteria may be linked to cravings.* 17 January 2008. Viewed 26 October 2010. <http://www.bodyecology.com/08/01/17/food_cravings.php>

Imaging reveals 2010. *Key metabolic factors of cannibalistic bacteria.* ScienceDaily, 7 September 2010. Viewed 6 February 2011. <http://www.sciencedaily.com/releases/2010/09/100903092513.htm>

Immunization 2009. *(Vaccination) is a way to trigger your immune... Vaccinations have been one of the most important health advances in history...*2 November 2009 viewed 8 February 2011 <http://health.nytimes.com - Times Health Guide>

Infection of the Body by Bacteria. *Why do bacteria invade the body...The main reason why bacteria invade the body of humans....* Viewed 20 February 2011 <http://crohn.ie/archive/primer/infect.htm>

Infectious diseases America's 3rd leading Killer: *What can you do? Astonishingly about 150.000 Americans die every year from taking medication as prescribed....* Viewed 4 December 2010 <http://www.pizzazz.net/new.page.2.htm>

Influenza, commonly referred to as the flue, is an infectious disease cause by RNA viruses of the family Orthomyxoviridae (the influenza viruses) that affects birds and mammals. Viewed 5 March 2011 <http://en.wikipedia.org/wiki/influenza>

In History, more soldiers have died from disease than from the enemy. Viewed 22 November 2010. <http://library.thinkquest.org/11170/epidemics/>

Inoculation is the placement of something that will grow or reproduce, and is most commonly used in respect of the introduction of a serum, vaccine, or antigenic substance...Viewed 8 February 2011. <http://en.wikipedia. org/wiki/Inoculation.commonly.used. in.respect.of....>

Intelligence of Cells. *To the best of my knowledge, the term CELL INTELLIGENCE was coined byNels...* Viewed 4 February 2011. <http://www.basic.northwestern.edu/g-buehler/summary. htm>

Interactions 2010. *Between human and microbial cells determine health ...grant to analyze ethical, policy problems linked with DNA practices* ... 11 June 2010. Viewed 15 February 2011 <http://www. news-medical.net/.../Interactions-between-human-and-microbial-cells-determinehealth-Hysical-well-being Researchers. aspx>

Isolation, 2010 *An Ancient and Lonely Practice, Endures...When patients turn in bed, giant waves of bacteria rise and travel on air currents all over the room,* ... Health Care Repeal Vote Scheduled by House Republicans... 31 August 2010. Viewed 12 January 2011. <http://wwwnytimes.com/2010/08/31/health/ 31essay.html>

Jacob, Ben Jacob; Becker, Israela; Shapira, Yoash; Levine, Herbert. 2004. *Bacterial linguistic communication and social intelligence....* 8 August 2004. Viewed 28 December 2010. <http:// doi.10.10.16/j.tim.2004.06.006>

Jarvis, WT. 2000. Colonic Irrigation. *Embalmers observed the petrification by bacteria (abnormal process within...Just as the ancient Egyptians did, health neurotics may temporarily ...a gap within which practices such as colonic irrigation can flourish* ... 17 December 2000. Viewed 6 December 2010. <http://www.ncahf.org/ articles/c-d/colonic.html>

Jensen, Bernard, D.C.; Ph.D. 1982. *Iridology. The Science and Practice in the Healing Arts*. Volume II. Bernard Jensen International, Escondido CA

Jensen, Bernard D.C.; Ph.D.2000. *Guide to Body Chemistry & Nutrition*. Keats Publishing, Los Angeles.

Johnson, Thomas J. - *A History of Biological warfare from 300 B.C.E.to the Present. What is a biological or bio-warfare? It is the use of biological pathogens (bacteria, viruses, fungi, and toxins derived from living organism to kill or....*Viewed 1 February 2011. <http://www.aarc.org/resources/biological/history.asp/2010/20_2_germ-warfare.html>

Kapnistos, Peter Fotis 2009. *Quantum Entanglement implies faster than light-speed interaction...*What happens to our friendly microbes when we die? ... February 2009. Viewed 3 February 2011. <http://www.americanchronicle.com/authors/view/3800>

Kapnistos, Peter Fotis 2009. Probiotics: *Only One-Tenth of your Body Dies.* Merely ten percent in your body are human cells. The remaining 90% are...28 November 2009. Viewed 3 February 2011. <http://www.americanchronicle.com/articles/view/130490>

Kids Exposed to Bacteria 2010. *Have Stronger Immune Systems....They suspect that children need contact with disease-causing microbes...*10 March 2010. Viewed 2 January 2011. <http://www.alignlife.com/articles/childhealth/Kids_Bacteria_infections.html>

Killing Germs. 2005. *In Hospitals, Air Ducts with Silver Based Coating Stay Germ-Free. Preventing hospital infections from such stubborn bugs as Staphylococcus aureus....* 1 September 2005. Viewed 15 February 2011. <http://www.sciencedaily.com/videos/2005/0910-killing_germs.htm>

Knight, R. 2009. *Unique Human Microbe Communities Have Wide Implications For Human Health.* 7 November 2009. Viewed 21 January 2011. <http://www.medicalnewstoday.com/articles/170114.php>

Khoruts, Alexander Dr. 2010. *How Microbes Defend and Define Us. Fr. Alexander Khoruts had run out of options. In 2008, Dr. Kharuts, a gastroenterologist at the University of Minnesota.....*12 July 2010. Viewed 17 July 2010. <http://www.nytimes.com/2010/07/13/science/13micro.html>

Knoll, A. 2007. *How did life begin? What are the origins of life....* Viewed 21 January 2011. <http://www.pbs.org/wgbh/nova/evolution/how-did-life-begin.htm>

Lam, Vincent M.D.; Lee, Colin M.D. 2006. *The Flu Pandemic and you.* Doubleday Canada, a Division of Random House of Canada Limited.

Landauer, Ellen, 2010. *Probiotics are beneficial bacteria that live primarily in our digestive system.Many people are familiar with yogurt, a food that ideally contains these kinds of beneficial.....*24 May 2010. Viewed 27 February 2011. <http://www.disabledworld.com/medical/alternative/probiotics/digestion.php>Probiotics for Healthy Digestion /a>

Lane, Craig - *The Journey of Digestion - Unknown Microbial Synergy. This will dry it out further and could cause carcinogenic combinations of toxic waste with...*Viewed 3 December 2010. <http://macrobiotics.co.uk /journeyofdigestion.htm>

Leaky gut syndrome negatively effects assimilation of food *...probiotic bacteria which help us attack and kill harmful microbes in the intestines* ...Viewed 19 February 2011. <http://www.icnr.org/.../colostrum-shown-to-help-significantly-with-intestinal-permeability-andirritable-bowel-syndrome.htm>l

Lee, Yuan Kun. 2008. *Who are we? Microbes, The Puppet Masters! Our genome is littered with scraps of DNA, termed junk DNA, that serve no apparent purpose*.....December 2008. Viewed 14 January 2011. <http://www.worldbooks.com/lifesci/7053.html>

Lederberg, Joshua. 2010. *Of Men And Microbes. Understanding SARS and Infectious Disease. The great scientific news that greeted this century was the campaign to decode the human genome*....21 April 2003. Viewed 29 December2010. <http:www.digitalnpq.org/global_Services /nobel%20laureates /04-21-03.html>

Lipton, Bruce Dr. 2010. *Startling New Discoveries in Biology Make Spontaneous Healing of Disease Possible*...Viewed 7 February 2011. <http://www.thetappingsolution.com/cmd,php?-Clk=4132601>

Look at the different definitions of intelligence, how is it measured ..... and the theories of general and multiple intelligences. Viewed 4 February 2011. <http://www.aboutintelligence.co.uk-Types of Intelligence.>

Lopez Alvarez. MJ. 2007. *Proteins in human milk. The human baby is born extremely immature....Breastfeed* Rev.2007 Mar: 15(1);5-16. Viewed 21 January 21. <http://www.ncbi.nlm.gov/pubmed/17424653>

Maczulak, A. 2010. *Allies and Enemies: How the World Depends on Bacteria*. Viewed 21 January 2011 <http://www.ftpress.com/store/product. aspx?isbn=0137015461>

Margulis, Lynn. 2006. *Bacterial Intelligence. Origin and Evolution of Life. Bacteria may not have brains, but they are intelligent. So says*.... 10 December 2006. Viewed 21 October 2010. <http://www.astrobio.net/interview/ 2111/bacterial-intelligence>

McGrath, K. PhD. 2006. *Nano Silver Kills Germs and Promotes Healing. Microorganisms have been around before human beings inhabited the planet. Many are necessary*....April 2006. Viewed 11 November 2010. <http://www.qsinano.com/news/newsletter/2006.../2006_04_f7.php->

Medicinal Uses of Silver. *Silver has a long and illustrious history of medicinal use. A brief history* ...Viewed 11 October 2010.<http://www.sterling-silver.ws/articles/about-silver/medicinal-uses-ofsilver.htm>

Merck Manual of Medical Information, Home Edition, 1999. Pocket Books, a division of Simon and Schuster, New York. NY, pages 883 - 889

Mercola, Dr. 2008. Kanzius Machine. *A Cancer Cure? John Kanzius, a man with no background in science or medicine, has come up with what may be one of the most promising breakthroughs....* Viewed 28 January 2011 <http://.www.mercola. com/js/citation.js language=javascript./script>

Mesoderm is found in most large, complex animals. Viewed 19 October 2013. <http://www.princeton.edu/-achaney/tmve/wiki100k/docs. Mesoderm.html>

Messaoudi,M, Lalonde R, Violle N, Javelot H, Desor D, Nejdi A, Bisson JF, Rougeot C, Pichelin M, Cazaubiel M, Cazaubiel JM. 2010. *Assessment of psychotropic-like properties of a probiotic formulation*....26 October 2010. Viewed on 16 January 2011. <http://www.ncbi.nlm.nih.gov/pubmed/20974015>

Microbe Composition 2009. *In Gut May Hold Key To One Cause of Obesity.* Science Daily. 20 January 2009. Viewed 11 December 2010. <http://www.sciencedaily.com/releases/2009/01/090119210437.htm>

Microbes are tiny organisms. 2010 -----*too tiny to see without a microscope, yet they*...Updated 21 July 2010. Viewed 21 November

2010. <http://www.niaid.nih.gov/topics/microbes/Pages/default.aspx>

Microbes have Consciousness. 2007. *Despite their lousy name, Slime Molds are among the most amazing organisms in existence...* 15 September 2007. Viewed 10 February 2011. <http://www.idscience .org/2007/09/15/microbes-have-consciousness-2/>

*Microbes have existed for millions, and even billions of years, their presence was not detected until the seventeenth century....*Viewed 2 December 2010. <http://faculty.cascadia.edu/jvanleer/astro%20sum01/goingtoextremes/ microbes.htm>

Microbes in the human body (MPKB). *Such interference results in genetic mutations, meaning that human DNA is almost.... ecology of interactions between human skin micro biota and mosquitoes...*Viewed 15 February 2011. <http://mpkb.org/home/pathologenesis/micrbiota>

Microbes live in symbiosis 2009. With our organism. ... *But, conventional medicine is still hanging on to Pasteur's ideas. ...*5 November 2009. Viewed 12 January 2011. <http://curezone.com/blogs/fm.asp?i = 1519006>

Microbes Symptoms. 2007. *...Some infectious diseases, such as the common cold, usually do not require a visit to your doctor. They often last a short time and are not....*Updated 3 November 2010. Viewed 16 December 2010. <http://www.niaid.nih.gov..Topics..Microbes..Understanding..>

Microbes Transmission. *According to healthcare experts, infectious disease caused by microbes....*Updated 3 November 2010. Viewed 4 February 2011. <http://niaid.nih.gov/topics/microbes/understanding/ transmission/pages/default.aspx>

Microbial consciousness, plausible? Philosophy discussion. Viewed 12 February 2011. <http://www.physicsforum.com-General--Discussion Philosophy>

Microbial Ecology is the relationship of microorganisms with.... modified 26 September 2010. Viewed 21 January 2011 <http://en.wikipedia.org/ wiki/Microbial_ecology>

Microbial intelligence (popularly known as bacterial intelligence) is the intelligence shown by microorganisms. The concept encompasses complex adaptive....Viewed 4 October 2010. <http://en.wikipedia.org/wiki/Microbial _intelligence>

Microbial metabolism. *Microbes use many different types of metabolic strategies and species can often be ....chemolithoautothrophs obtain energy from the oxidation of inorganic...* modified 4 February. Viewed 19 February 2011. <http://en.wikipedia.org/wiki/Microbial_metabolism>

Microbiology - The Study of Microbes.... *For example some species of Clostridium are able to digest carbohydrates and make alcohols and a very smelly organic ...* Viewed 2 February 2011 <http://www.disknet.com /indiana_biolab/b003.htm>

Microorganism *(from the Greek: mikros, small and organismos, also spelt microorganism) or microbe is an organism.....* Viewed 21 September 2010. <http://en.wikipedia.org/wiki/Microorganism>

Miller, Judith - *When germ warfare happened* by Judith Miller, City Journal Spring 2010. Seventy years ago,Japan's bio-attacks killed hundreds of thousands. The effects linger today. Spring 2010. Viewed 1 February 2011. <http://www.city-journal.org>

Mitchell, Stewart. B.Phil. 1998. Naturopathy, *Understanding the Healing Power of Nature. Our thoughts and feelings are intimately connected to the workings and maintenance of our body....the susceptibility of a person to succumb to disease seems to depend on an inner personal profile*. Element Books Limited. Shaftesbury, Dorset, Great Britain; Boston, Massachusetts, United States; Ringwood, Victoria, Australia pages 14-15

Mitchum, Robert; Tsouderos, Trine. 2009. Virus invasion sets off battle inside the body...Like a sleeper agent, the flu virus cause its damage from within, turning.....1 May 2009. Viewed 20 February 2011. <http://physorg.com/news160411270.html>

Montenegro, Maywa. 2010. *Petri Dish Images of Bacterial Colonies Show The Complex patterns That Emerge As Bacteria Cope in A Hostile Environment*....25 march 2010. Viewed 28 December 2010. <http://seedmagazine.com/content/article/portfolio_colonial_intelligence/>

Munger, Dave. 2009. *Antibiotic resistance is more than just a medical scourge; it s also a window into a war microbes have been waging*...7 October 2009. Viewed 3 January 2011. <http://seedmagazine.com/article_warfare/>

Mutating bacteria challenge 1994. *Medical science has made amazing advances in this century in combatting disease-causing microbes. But bacteria*....Discover, August 1994, p. 45

Naturally Healthy, 2007. BIOLOGICAL TERRAIN VS GERM THEORY. *The germ - or microbian - theory of disease was popularized by Louis Pasteur (1822-1895), the inventor of pasteurization. This theory*...29 October 2007. Viewed 28 December 2010. <http://timelessremedies.wordpress.com/.../biological-terrain-vs-the-germ-theory/->

Nayreet, Sly. 2006 A Brief History of Antibiotics. *We haven't always relied on the latest new medicines to remedy what ails us, we used what we had around us*...15 December 2006. Viewed 12 January 2011. <http://www. associatedcontent.com/article/100649/a_brief_history_of_antibiotics_pg2.html>

Newman, J. Dr. 1995. *How Breast milk Protects Newborns. Doctors have long known....* Viewed 21 January2011. <http://sciamdigital.com/index.cfm?fa>

New viruses and bacteria threaten1994. *New viruses and bacteria threaten. New strains of deadly viruses and old, familiar bacteria with newly gained resistance to antibiotic drugs.....*TIME, Sep. 12, 1994, p. 62-69

Oddy, WH. 2001. *Breast-feeding protects against illness and infection in infants and children: a review of the evidence.* Breastfeed Rev. 2001 Jul; 9(2): 11-8 Viewed 21 January 2011. <http://www.ncbi.nlm. nih.gov/pubmed/11550600>

Oddy, WH. 2002. *Long-term health outcomes and mechanisms associated with breastfeeding. Expert Rev Pharmacoecon outcomes* res. 2002 Apr; 2(2); 161-77. Viewed 21 January 2011. <http://www.ncbi.nhm.nih.gov/ pubmed/19807327>

Oddy, WH. 2002. *The impact of breast milk on infant and child health.* Breastfeed Rev.2002 Nov, 10(3):5-18: Viewed 21 January 2011. <http:www.ncbi.nlm.nih.gov/pubmed/12592775>

Omnipotent Grace. 2007. *When did evolution start?* Viewed 17 December 2010. <http://omnipotentgrace. wordpress.com/2007/01/26/when-did-evolution-start/>

One Microbes Meat. 1995. *Newly discovered bacteria, dubbed MIT-13, use arsenic in their energy generation process in much the same way that humans use oxygen.....*March, 1995, p 20.

Origins of Life on Earth. 2000. *In general, organisms over time in the evolutionary chain have grown and become more complex in their nature, i.e. the first origins of life...*viewed 5 January 2011. http://biology-online.org. Tutorials The origins of Life>

Origins of Life on Earth. 2000. *Looking at how the geological change of Earth over time was able to support human life starting billions of years ago.* Viewed 8 December 2010. <http://www.biologyonline.org. Tutorials... The Origins of Life>

Outsmarting Killer Bacteria. 2010. *Antibiotics can work miracles, knocking out common infections like bronchitis and tonsillitis. But according to the Center for Disease Control, each year 90,000 people in the U.S. die of drug-resistant superbugs* ....15 September 2010. Viewed 17 December 2010. <http://www.sciencedaily.com/releases/ 2010/09/100914131004.htm>

*Paenicaccillis Vortex is a species of pattern-forming bacteria, first discovered.....*Viewed 19 October 2013. <http:// en.wikipedia.org/wiki/ Paenibacillus_vortex>

Pierce J. Ph.D. 2006. The Brain. *The Owner's Manual For The Brain*. March 2006. Bard Press, Austin, Texas pg. 781

Poi, Fermentation. 2009, *poi, fermentation and Traditional Poi Cultures....during the ancient practice of fermentation. ...beneficial effects of fermentation and probiotic bacteria. ... poi appears to have a beneficial effect on health and therefore meets the definition of a probiotic.* ...18 September 2009. Viewed 12 November 2010. <http://www.poico.com/artman/publish/article_72.php>

Port, Tami. 2009. *Difference between Aerobic and Anaerobic Bacteria: Microbes. Their Metabolism and relationship with oxygen.* 15 July 2009. Viewed 27 October 2010. <http://www.suite101.com/.../difference-between-aerobic -anaerobic-bacteria-a132294>

Port, Tami. 2010. *What is Microbiology? The Science of Microscopic Organisms - Studying Microbes - Microbiology, the study of microscopic (very small) forms of life...*1 April 2010. Viewed 11 November 2010. <http://suite101.com/content/what-is-microbiology-a220778>

Postgate, John, 2000. 4th Edition. *Microbes and Man.* Cambridge University Press, Cambridge, United Kingdom. Pages 133 - 138

Press, Simone - *Eat Poop? Why Docs Say It Can Be a Good Idea. These days, one of the hottest trends in transplant medicine is fecal flora...* A new study shows that even a single...Viewed 6 February 2011. CBS News

Probiotic Bacteria and Your Health. *There are many health benefits of probiotics. The fact of the matter... This includes our physical and mental health, our metabolism, and quite possible...*Viewed 28 December 2010. <http://www.globalhealingcenter.com -... - The Benefits Of...>

*Probiotics provide sustainable options for improving agricultural / environments performance. All living systems - including soil, plants, and trees - have a microbial....*Viewed 3 February 2011. <http://www.scdprobiotics. com/Agriculture's/311.htm>

Purdy, Michael. 2006. *Gut microbes, partnership helps body extract energy from food....*12 June 2006. Viewed 16 January 2011. <http://news.wustl.edu/news/Pages/7328.aspx>

Quantum Education - Where health matters! In 1979, Dr. Hamer began his research after the death of his son Dirk....The Ontogenetic System of Microbes. -microbial activity correlating with the...Viewed 16 February 2011. <http://www.quantumeducation.comgermannewmedicine.html>

Quantum suicide and immortality. *In quantum mechanics, quantum suicide is a thought experiment. It was originally published independent....* Viewed 16 February 2011. <http://wikipedia.org/wiki/Quantum _suicide_and- Immortality>

Quarantine (divine) at dictionary.com. Quarantine qua-rantine (kwor-en-ten). *A period of time during which a...quarantine.*

Viewed 1 February 2011. <http://dictionary.reference.com/browse/quarantine>

Quorum sensing. 2007. *What the math makes clear is that the purpose of quorum sensing is not a conscious attempt on the part of bacteria to coordinate behaviour.* ...24 October 2007. Viewed 8 December 2010. <http://wikidoc.org/index.php/Quorum_sensing>

Quorum Sensing in bacteria mirrors peaceful conquest behaviours. *Quorum sensing is a kin recognition mechanism which bacteria use to implement density...* Viewed 18 December 2010 <http://majorityrights.com/index.php/weblog/comments/quorum_sensing_in_bacteria_mirrors_peaceful_conquest_behaviour>

Raffa, Robert B; Iannuzzo, Joseph R; Levine, Diana R; Saeid, Kamal K. Bacterial Communication *(Quorum Sensing) via Ligands and Receptors; A Novel Pharmacologic target for the Design of Antibiotic Drugs.* Viewed 24 January 2011. <http://jpet.aspetjournals.org.doi:10.11.24/jpet.104.075150>.

Rawlings, Deidre PhD ND MH CNC *Root Canal Cover Up Exposed.* An interview with Dr. George Manning, D.D.S. Dr. Manning - Brings a most curious perspective to an expose of the latent dangers..... Viewed 29 December 2010. <http://www.selfgrowth.com/artivles/root_canal_cover_up_exposed>

Regeneration (biology). *Every species of regeneration, from bacteria to humans. .... Humans Could Regenerate Tissue like Newts by Switching off a Single Gene...*last modified 5 February 2011. Viewed 19 February 2011. <http://en.wikipedia.org/wiki.Regeneration (biology)>

Roman Medicine. *Ancient Roman medicine was a combination of some limited scientific........ Aesculapius, the god of healing, was the prominent deity that governed the Roman medical practices and his ... Roman society maintained reasonably good health throughout its history. ... And regular cleansing helped fight germs and bacteria. ...* Viewed 16 November 2010. <http://www.unrv.com/culture/roman-medicine.php>

Roth, Ronald, D.Ac – *Natural Common Cold and Flu Remedies for individuals with a normal and compromised immune system.* Viewed October 20, 2013. <http://www.acu-cell.com/flu.html>

Royal Rife, a pioneer in Energetic-Medicine Research. *In 1920 Royal Rife identified the human cancer virus using the world's most powerful microscope...*Viewed 2 January 2011. <http://quantumworld.info/Roya l%20Rife%20 Research.html>

Ru, Pravda. 2010. *Bacteria Rule the World, Starting with Humans. The humans inside look like a virtual zoo, full of a wide variety of bacteria, a new study found....*4 March 2010. Viewed 16 January 2011. <http://english.pravda.ru>News>Health>

Sagan, Carl and Druyan, Ann. 2007. *Consciousness comes from DNA...attention to every pathetic little microbe when they already know perfectly...Souls and consciousness could then pass, on their own ...* May 11, 2007. Viewed 11 February 2011. <http://richarddawkins.net/articles/1057>

Sahley, Bill J. PhD., C.N.C; Birkner, Katherine M. C.R.N.A., PhD; 2005. *Heal with Amino Acids and Nutrients. Survive Stress / Anxiety, Pain, Depression,* What to Use and When. Pain and Stress Center, San Antonio, TX pages 162-163

Sampson, Tony D. 2009. Dr. Thacker's Position Paper. *Surely we human beings are more than the microbes that inhabit us. Microbes strictly speaking...* 11 June 11 2009. Viewed 11 December 2010. <http://www.networkpolitics.org/request-for.../dr-thackers-position-paper>

Say Hello to the Bugs. 2007. *Say Hello to the Bugs in Your Gut. Your small and large intestines are home to countless microbes that scientists think*...31 December 2007. Viewed 11 December 2010. <http://www.newsweek.com /2007/.../ say-hello-to-the-bugs-in-your-gut.html>

Schirber, Michael. 2006. How Life began; *New research Suggests Simple Approach. Somewhere on Earth, close to 4 billion years ago...* 9 June 2006. Viewed 26 December 2010. http://www.livescience. com animals/060609_life_origin.html

7 Worst Pandemics. 2009. 7 Worst Pandemics in History. *A pandemic ia an epidemic of infections*...27 April 2009. Viewed 28 October 2010. <http://www.impactlab.net/2009/04/27/ 7worst-pandemics-in-history>

Shekhar, Chandra. 2008. Thinking ahead: *Bacteria anticipate coming changes in their environment. Microbes may be smarter than we think. A new study by Princeton University researcher's shows.* 9 June 2008. Viewed 3 December 2010.<http://www.molbio. princeton.edu>

Shomon, Mary. 2006. Probiotics *the Good Bacteria Could Be a Key to Good Health. Probiotics are bacteria we eat and they're good for our health*...5 March 2006. Viewed 14 *December 2010. <http://Thyroid Disease Blog. By Mary Shomon. Thyroid* about.com/.../probiotics-the-good-bacteriacould-be-a-key-to-good-health.htm>

Slade, SB; Schwartz, SA. 1987. *Mucosal Immunity of Breast milk.* In Journal of Allergy and Clinical Immunology, Vol.8, No. 3, pages 348-356; September 1987

Sleator AD. 2009. *The human super organism - of microbes and men.* 16 October 2009. Viewed 20 January 2011. <http://www.medical-hypotheses.com/article/S0306-9877(09)00658-6/abstract>

Smith, Gerald H. D.D.S., N.M.D. *The Oral Connection. Oral Pathology Cover-Up Exposed. Conventional dental procedures offer* *a technique which does not...*Viewed 21 November 2010 <http://www.Newhopetechnologies.com/oralconnection.htm>

Smith, Leonard Dr. *Is It Possible to Get Too MUCH FERMENTED FOOD IN YOUR DIET? Experiencing gas and bloating or other symptoms when you consume fermented foods...*Viewed 11 December 2011. <http://www.bodyecology.com/07/05/24/too_much_fermented_food.php>

Smith, Tara C. Assistant Prof. 2007. *Fecal transplants to cure Clostridium difficile infection.* In my field, many things that cause the average man-an-the-street to get a bit squeamish are rather commonplace...17 December 2007. Viewed 6 February 2011. <http://scienceblogs.com/.../12fecal_transplants_to_cure_clos,php>

Spreen, Allan, M.D. 2009. *Tomorrow's Cancer Cures Today.* Health Science Institute, Maryland

Squatting - 2010.... *So for speedy, efficient, effortless and enormous bowel movements, try squatting because squatters do it better! Last but not least....* Viewed 8 January 2011. <http://www.colonics-brisbane.com.au /SQUATTING>

Sreeraman. 2009. *Might of the Microbes. Several of these tiny creatures called micro-organisms languish in anonymity and are little known.* But they are.... 18 August 2009. Viewed 4 December 2010. <http://www.medindia.net/news/healthwatch/might-of-the-Microbes-56504-1.htm>

Stern, Alexandra Mina. 2005. *The History of Vaccines and Immunization: Familiar Patterns, New Challenges* -Markel, Howard - Human beings have benefited..... Viewed 8 February 2011. <http://content.healthaffairs. org/content/24/3/611.full>

Stomach Enzymes, Digestive Enzymes, *Digestive Health Supplements. Cellulase- is made from microbes, and breaks down fiber, which would be classed as a carbohydrate, although without this enzyme we are unable to digest it...* Viewed 2 February 2011. <http://www.effective-healthsupplements.com/Stomach-Enzymes.html>

Stop Sugar Cravings on HCG Diet. *Sugar cravings seem to be worse while on the HCG diet. That is...*Viewed on 4 December 2010. <http://besthcgweightloss.comstop-sugar-cravings-on-hcg-diet/>

The Academy of Traditional Chinese Medicine. *An Outline of Chinese Acupuncture*. Foreign Language Press, Peking 1975

The Microbe Zoo. 2010. DLC-ME / The Microbe Zoo / Information Booth / *what is a Microbe? A microbe is any living organism that spends its life at a size too tiny....*11 September 2010. Viewed 5 February 2011. <http://microbezoo.commtechlab.msu.edu/zoo/>

The Origins of Life. 2000. Origins of Life on Earth. - *The origins of Life. Although Earth was crated around 4.5 billion years ago, life began to exist not long after...* 1 January 2000. Viewed 26 January 2011. <http://www.biology-online,org - Tutorials - The origins of Life>

The Silver Spoon. *Feed your baby with a Silver Spoon. Most people know the saying Born with a silver spoon In your mouth. This means that the child....*Viewed 4 December 2010. <http://hubpages.com/hub/Feed-your-Baby-with-a Silver-Spoon->

*These microbes are a natural part of our system that are established from the breast milk you receive....*Viewed 7 November 2010. <http://www.nutrinexus.com/about-nutrinexus/>

Thinking Microbes / Bite size Bio. 2007. *Cognition is a term frequently used in several loosely related ways to refer to a faculty for the-human-like processing of information....*Viewed 3 December 2010. <http://bitesizebio.com2007/12/10/ thinking-microbes/>

Todar, Kenneth, PhD. 2008. *The Normal Bacterial Flora In a healthy animal, the internal tissues,... The predominant Bacterial flora of humans are shown...* Viewed 27 November 2010. <http://www. textbookofbacteriology.net /normalflora.html>

Todar, Kenneth, PhD 2011. *The Impact of Microbes on the Environment and Human Activities.* Microbes are everywhere in the biosphere, and their presence invariably affects ... Viewed 16 February 2011. <http://textbookof bacteriology.net/Impact.html>

Torkos, Sherry, Bsc, Phm and Wassef, Fraid, Rph, CCN. 2003. *Breaking the Age Barrier*. Strategies for Optimum Health, Energy and Longevity. Penguin Group, Toronto, Ontario. Page 70

Toward A. Rosetta Stone. For Microbes Secret Language. 2007. *Scientists are on the verge of decoding the special chemical language that bacteria use*....31 December 2007. Viewed 12 December 2011. <http://www.sciencedaily.com/ releases/2007/12/0710094701.htm>

True Colloidal Silver. *Silver has been known for its health values since before the time of Julius Caesar. Romans used*....Viewed 4 December 2010. < http://ultimate.org/silver/ >

Tylec, J. 2007. *Babies and Friendly Bacteria. Within a few days of birth your baby s digestive tract*. 2 August 2007. Viewed 23 January 2011. <http://ezinearticles.com/?Babies-and-Friendly-Bacteria&id=672129>

Tyson, Peter. Nova / The most Dangerous Women in America / History of Quarantine...*Explore this Time line of history of quarantine, the practices of segregating the sick or exposed from the healthy, which stretches back to Roman times*. Viewed 2 February 2011. <http://www. Pbs.org/wgbh/nova/typhoid/ quarantine.html>

Unexpected Microbe Diversity on Human Skin. 2009. *A new study of the skin s micro biome----All of the DNA, or genomes, of all of the microbes that*....Viewed 29 October 2010. <http: / / www.nih. gov...-Archive June2009Archive>

Vaccination is the administration of antigenic material (a vaccine) to simulate adaptive immunity to a disease. Vaccines can prevent or ameliorate the....Viewed 8 February 2011. <http: / / en.wikipedia.org / wiki / Vaccination.>

Van Over, Raymond. I Ching – A Mentor Book. New American Library. New York NY

Vertebrate endoderm development and organ formation. Viewed 19 October 2013. <http: / / www.ncbi. nlm.nih.gov / pubmed / 19575677>

Viral phenomenon: 2010 *Ancient microbe invade human DNA. Humans carry in their genome the relics of an animal virus that infected their forerunners at least 40 million years ago*...6 January 2010. Viewed 12 January 2011. <http: / / www.physorg.com / news182006019.html>

Virology Journal 2010, 7:42doi:10.1186/1743-422X-7-42 Published 18 February 2010

War on bacteria is Wrongheaded. 2006. *Children not exposed to harmful bacteria or conversely, given antibiotics to kill bacteria, do not receive the germ workout required to*...28 March 2006. Viewed 2 January 2011. <http: / / www.livescience.com / health / 060328_ bad_bacterial.html>

Washington, Harriet. Virus and mental *illness. Microbes that cause mental illness can also enter the body another way on one fork. In the mid-1990s, an outbreak of Creutzfeldt-Jacob disease struck fear* ... Viewed 4 January 2011. <http:www.vaccinetruth.org / virus_ and_mental_illness_.htm>

Wassenaar, T.M. Dr. 2008. Bacterial Disease History. *Bacteria existed long before humans evolved, and bacterial diseases* ...Updated 1 March 2009. Viewed 21 November 2010. <http://www.bacteria museum.org/cms/ Special-features/bacterial-diseases-in-history.html>

Weintraub, Skye N.D. 2000. *Parasite Menace*. Woodland Publishing, Pleasant Grove, Utah US. Pg. 4349

Wenner, Melinda. 2008. Infected with Insanity: Could Microbes Cause Mental Illness ... And yet studies have repeatedly linked schizophrenia to prenatal infections with influenza virus and other microbes, showing that the ... 17 April 2008. Viewed 6 January 2011. <http://www.scientificamerican.com/article.cfm?id=infected-with...>

What Is Biological Terrain And Why Care? *The human body literally hosts a universe of microbes (Kroes, 1999). It is the bio-terrain - the internal milieu/environment....* Viewed 14 December 2010 <http://www.healinggrapevine.com/.../what-is-biological-terrain-and-why-care.html>

What is Intelligence? *The article will look at how society measures intelligence,...*Viewed 4 February 2011. <http://hubpages.com/hub/ Psychology-101-What-Is-Intelligence>

What Is Life? *Early Life Over a very long time, gradual changes in the earliest cell...*Viewed 4January 2011. <http://windows 2universe.org/the National Earth ScienceTeachers.www.windows2universe.org/Life/life1.html>

What is Microbiology? *What the Heck is Microbiology? Microbiology is the study of microorganisms little, really little, critters (except for The Big One)...*Viewed 21 November 2010 <http://People.ku.edu/jbrown/whatmicro. html>

What is Quantum AG? *Quantum AG is simply a recognition of the nature of Nature putting this knowledge to work. All other...* Viewed 16 February 2011. <http://www.cropserviceintl.com/Technology/quantum.ag.htm>

Where and when did life begin? 2000, origins of Life on Earth. Where and when did life begin? Discusses microbes on early earth and how life evolved from early microorganisms. Viewed 5 December 2010. <http://www.space.com/searchforlife/life_origins_001205. html>

Where did life begin? 1993, *Microbiologists, geochemists, astronomers and oceanographers are researching the origin of life. A microbial community was already....*Time, October 11, 1993, p 68-74

Whiteman, L. 2008. *Nowhere is the principle of strength in numbers....*Viewed 21 January 2011. <http://www.usnews.com/science/articles/2008/04/08/ microbes-to-people-without-us-yourenothing>

Why silver is your families most precious metal. *Nothing matters as much as your family's time together. Really making most....*Viewed 11 December 2010. <http://www.elastoplastsilverhealing.co.uk/>

Why the Arsenic-Eating Microbe is a Huge Deal. 2010. *NASA has found a new form of life?* 2 December 2010. Viewed 11 December 2010. <http://www.cbsnews.com/stories/2010/12/02tech/main7111270.shtml>

Williams, Paul. 2007. *Quorum sensing, communication and cross-kingdom signalling in the bacterial world.* Viewed 28 December 2010. <http://www.ncbi.nlm.nih.gov/pubmed/18048907>
Wing, Thomas W. Dr. 1976. *The Theory of Applied Electro Acupuncture and Technic of Non-Needle Electro-Acupoint Therapy.* SEE-DO Press, Pomona CA.

Wing, Thomas W. Dr. 1976. *The Theory of Applied Electro Acupuncture and Technic of Non-Needle Electro-Acupoint Therapy.* SEE-DO Press, Pomona CA.

Zink, Albert PhD. Ancient Egypt Medicine. *The practice of medicine in ancient Egypt, the physicians, their instruments and medicines. ... In Sickness and in Health. Preventative and Curative health Care.....bacteria in an infant Mummy from Ancient Egypt* by Albert Zink PhD, Viewed 12 January 2011. <http://www.reshafim.org.il/ad/Egypt/timelines/topics/medicine.htm>

Zugler, Abigail MD. 2010. Isolation, an Ancient and Lonely Practice, Endures....*For those who are not just infected on the inside but also infested on the outside....*Viewed 5 February 2011. <http://www.nytimes.com/2010/08/31/health/31essay.html>

CPSIA information can be obtained at www.ICGtesting.com
Printed in the USA
LVOW06s0241110314

376803LV00011B/54/P